SIMON & SCHUSTER

NEW YORK LONDON TORONTO

SYDNEY TOKYO SINGAPORE

DAVID FREEDMAN

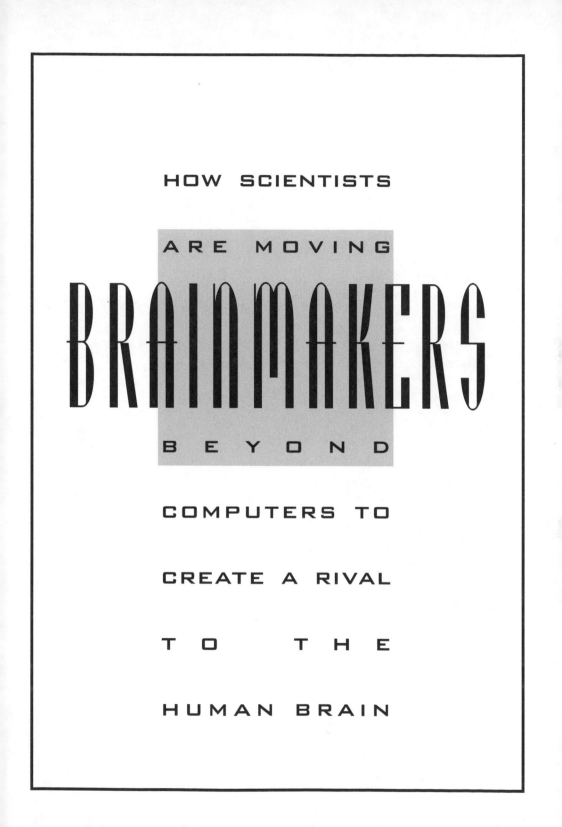

HOW SCIENTISTS

ARE MOVING

BRAINMAKERS

BEYOND

COMPUTERS TO

CREATE A RIVAL

TO THE

HUMAN BRAIN

SIMON & SCHUSTER
Rockefeller Center
1230 Avenue of the Americas
New York, New York 10020

Designed by Songhee Kim
Manufactured in the United States of America

10 9 8 7 6 5 4 3 2 1

Library of Congress Cataloging in Publication Data

Freedman, David, date.
    Brainmakers: how scientists are moving beyond computers to
create a rival to the human brain/David Freedman.
        p.   cm.
    Includes bibliographical references and index.
    1. Artificial intelligence.   2. Brain.   I. Title.
Q335.F735   1994
006.3—dc20   93-33632
                                                  CIP
ISBN: 0-671-76079-3

ACKNOWLEDGMENTS

I'm grateful to Paul Hoffman, Will Hively, and the other *Discover* magazine editors for the assignments that inspired this book. Thanks to Fahria Rabbi for research assistance, Mike Freedman for careful reading, Laurie for ideas and encouragement, and Rachel and Alex for strategic distraction. Finally, I'm indebted to Jane Dystel, my agent, for support and excellent advice, and to my editor, Bob Asahina, for his commitment and direction.

To my parents

CONTENTS

We are here for this—to make mistakes and to correct ourselves, to withstand the blows and to hand them out. We must never feel disarmed: nature is immense and complex, but it is not impermeable to the intelligence; we must circle around it, pierce and probe it, look for the opening or make it.

—Primo Levi

INTRODUCTION

The designers of the first flying machines took as their inspiration the flight of birds. But the flapping contraptions that resulted failed, and by the middle of the eighteenth century most inventors of flying machines had turned their attention away from birds and toward the mathematical expressions that embodied nascent aerodynamic theory. By the turn of the century the airplane had made its debut.

Nature had provided a sort of blueprint for flight in the form of the bird, but it was a blueprint that turned out to be more opaque than the underlying principles of flight. The achievement of the airplane was just one of many triumphs of the period that seemed to suggest that science's ability to analyze its way around nature was nearly limitless. The steam and internal combustion engines, based on the new science of thermodynamics, were transforming society. French mathematicians Pierre-Simon Laplace and Jules-Henri Poincaré had refined Newtonian physics to the point where not only the motions of every body in the solar system but theoretically of every particle in the universe could supposedly be predicted with perfect accuracy, given enough initial information. Engineer-turned-manager Frederick Winslow Taylor carried this same spirit of analysis, control, and predictability to the workplace, where equations describing the lifting capacity of human beings were used to calculate maximally efficient work routines.

The greater part of the twentieth century saw little reason to question this approach, as the hard sciences turned out a stream of successful theories and applications, reaching a zenith of sorts with the Manhattan Project. Biology may have been buried in a morass of mysteries and complexity, but physics and chemistry seemed capable of analyzing almost any phenomenon, expressing its underlying principles as equations, and then creating it from scratch in a laboratory.

It was in this cocky intellectual atmosphere that the new science of artificial intelligence was born in the 1950s. AI seemed destined to sail through the same process of distilling first principles and then applying them in a straightforward fashion. Everyone in the field seemed convinced that intelligence would yield its secrets in no more time than it had taken the atom to do the same.

But everyone in the field was wrong. Unlike flight or the nucleus of the atom, intelligence proved too difficult to figure out. This is the story of how a promising young science ran aground, and of a new movement that has emerged to carry it forward by stepping back to take another look at how birds fly.

BRAINMAKERS

1. BUG BRAINS

I will confess that I have no more sense of what goes on in the mind of mankind than I have for the mind of an ant. Come to think of it, this might be a good place to start.

—LEWIS THOMAS

I have developed a great respect for the engineering that went into the visual systems of animals. I always find myself saying, "I never would have thought of that, but that's a good idea."

—CARVER MEAD

The MIT Artificial Intelligence Lab could pass at times for the world's largest and best-equipped playroom. Among the goodies lying in and around the common area are a large plastic dinosaur, a toy battlefield complete with tanks, a jar of what looks like purple phlegm, and an outsized blackboard on which the following has been carefully written:

TOFU WEENIES
BEEF WEENIES
KOSHER WEENIES (BEEF)
TECHNO WEENIES (GEEKS)

An even more juvenile ambience can be encountered one floor above: a ten-foot-square Plexiglas-enclosed sandbox, complete with toy bulldozers, a two-foot-long plastic ant, a ten-inch-long metal cockroach, and a fortyish, cherubic-looking scientist who could be mistaken at first glance for a really tall kid at play. Though it is not quite as large as the plastic ant, the metal cockroach proves the most interesting of the toys: it crawls, eerily insect-like and with uncanny dexterity, across the sandy landscape around it.

The scientist is Rodney Brooks, and the cockroach is Attila, Brooks's most advanced mobile robot. Attila doesn't do much of anything besides crawl around at one and one-half miles per hour and try not to bump into things. But appearances are deceiving. "Ounce for ounce, Attila is the world's most complex robot," beams Brooks. "Maybe in absolute terms, too." Attila's 3.6-pound, six-legged frame carries 24 motors, 10 computers, and 150 sensors, including a miniature video camera. Each leg has four independent ranges of motion, allowing it to climb over objects, scramble up near-vertical inclines, and pull itself up onto ten-inch-high ledges.

More important, Attila incorporates "the subsumption architecture," Brooks's unique approach to the software that controls the robots. AI has always held that achieving useful behavior in a robot required complex programs that exerted tight control over every possible action, and yet more complex programs to keep track of its surroundings and negotiate them according to preset routines. Brooks, on the other hand, wants to bury that tradition and start from scratch with a new notion of the nature of intelligence, one that is, at least in its initial incarnation, closer to the spinal cord than to the brain. Instead of employing a "top-down" approach of explicitly programming in intelligence, Brooks insists that intelligence emerge on its own, in a "bottom-up" fashion, through the interaction of relatively simple independent elements—as it apparently does in nature.

This sharp departure from tradition has created something of a fracas in the AI community, and Brooks, the self-styled Bad Boy of robotics, is doing everything he can to exacerbate it. "There's a deep difference in philosophy right now about how to approach artificial intelligence," he says. "All the other guys are wrong, but they're welcome to go out and do it their way if they want."

Unlike industrial robots that repeat the same action over and over again, an autonomous robot would be able to adapt its behaviors to the demands of each situation, or even choose from among a repertoire of widely varying actions. Besides embodying one of the most powerful and enduring of human fantasies, highly functional, autonomous robots would also create a market likely to dwarf that of the $500 million annual demand for assembly-line-style robots.

Since choosing appropriate behaviors requires a certain minimal intelligence, autonomous robotics is considered a subfield of AI. But robotics adds an entirely new challenge to the already formidable challenges of reasoning: a robot must not only come up with an answer to a problem, it must embody this answer in the real world, where even small errors can render a robot useless or even dangerous.

The task of building even the simplest autonomous robot proved so difficult to AI's pioneers that by the late 1970s researchers had scaled back their ambitions to achieving particular, narrowly defined robot capabilities. Large parts of careers have been spent on developing robot grasping, robot walking, even robot juggling. Each task presents its own unique set of problems, and for each problem there is a range of proposed solutions, none of which ever quite seem to do the job.

One of the most sought-after, and challenging, capabilities is navigation. In other words, how can a mobile robot make its way across a room or a field without bumping into things? The standard approach to this problem is to provide the robot control program with some sort of conceptual model of the world. The software might incorporate a map of a room, for example, or a list of the characteristics that distinguish a chair from a table. When trying to navigate around a room, a robot equipped with such a program might spot an object, compare it to the various lists of objects' characteristics to identify it, locate the position of the object on the map, and then use this information to determine the direction in which it should head. The ability to recognize obstacles and follow internal maps has always been considered the starting points for mobile robots, the minimum qualifications for navigational intelligence.

For many years, this analytical approach to robot mobility was epitomized by Shakey, a state-of-the-art robot designed in 1969 by Nils Nilsson and others at Stanford. The five-foot-tall automaton's program, which was based on formal logic, required so much computing power (for its time) that it had to be provided by a separate computer attached to Shakey via cable. For all this, Shakey moved through a nearly empty room at the blinding speed of one foot every five minutes.

By the early 1980s mobile robots had become far more sophisticated and flexible, though they still relied on maps and a step-by-step problem-solving approach to navigation. One of the most advanced mobile robots of this type was the autonomous outdoor vehicle funded by the military and designed by David Payton and others at Martin-Marietta. The van-like vehicle, which was designed to make its way both along roads and across off-road terrain, was programmed with a map that divided the surrounding territory into squares; for each square, the program noted the elevation, soil hardness, vegetation, and other characteristics of the ground at that spot.

Every half second the vehicle's computer determined where it was and calculated the fastest path to its goal. Unlike Shakey, the vehicle carried its computing power on-board, but at a cost: the vast array of computer equipment was so heavy that the overburdened vehicle sagged to a mere six inches off the ground, preventing it from tackling all but the smoothest terrains. Even so, the vehicle traveled at a top speed of two miles an hour, and often got hopelessly off course within a few hundred feet. "The problem was that the map didn't have all the information it needed to make a perfect path," says Payton, noting that the vehicle often got lost or took circuitous routes the first time it encountered any obstacle its program hadn't anticipated. "It's like a grocery list. If you had to follow your list explicitly, without knowing about the kinds of trade-offs you have to make in sizes or brands, you'd never come back from the store."

Carnegie-Mellon scientist Reid Simmons ran into similar problems with Ambler, an eighteen-foot-tall, six-legged machine originally intended as a Mars explorer. To keep it from getting tripped up or bogged down, Simmons turned to heuristics, gathering together rules about the real world that the robot could employ in unexpected situations. "We know that if you step in something soft your foot will sink, or that if you put your foot down and you don't feel anything that the ground's not there," he explains. "We're trying to get robots to expect these sorts of things." Simmons has had to gather these rules and add them to the robot's database, but he is working on a program that he hopes will allow the robot to analyze its own mistakes and create new rules from them.

Despite such ambitious projects, most researchers concede that the progress of navigational robotics has been disappointing over the last twenty years. Perhaps symbolizing the frustrations of the field was the fate of Dante, the six-foot-tall, spider-like robot built by a Carnegie-Mellon team led by William "Red" Whittaker, a highly respected robot researcher.

Dante is a multimillion-dollar machine commissioned by NASA to descend into the mouth of Cerberus, an active volcano in Antarctica. After having spent two years working on Dante and its control program, Whittaker—who is only slightly less physically intimidating than his robots—philosophized about the machine's prospects the day before he and his crew were to ship out to Antarctica. "I realize there's every chance this is going to be a cataclysmic failure," he said. "I'm used to problems. There isn't a robot I've built that hasn't cut, bruised, or shocked me. But I think we're really prepared for this one." The next morning, a last trial run in a field near Whittaker's lab resulted in Dante shearing off four of its legs, as its control program failed to compensate for the legs' awkward positioning as they climbed over some rocks. Two months later, Whittaker and the team lowered a refurbished Dante sixty-five feet into Cerberus, only to have to bring it back out when the cable connecting Dante to its support equipment above was damaged. Dante has yet to fulfill its mission.

The large costs, long lead times, and generally unimpressive results of conventional AI approaches to autonomous robotics have led some investigators to question the field's basic approaches. In particular, many scientists have wondered over the years if roboticists wouldn't do better to take a closer look at how living creatures succeed so well at the same tasks at which robots seem so hopelessly inept. And where animals seem to excel compared to robots is in the realm of "low-level" intelligence.

"High-level" intelligence is what most of us think of when we consider our own mind's activities. It includes commonsense reasoning, problem solving, and creativity. It is also, according to a number of AI researchers, vastly overrated. The more important capabilities, they say, are the ones we take for granted: perception, motor control, reflexes. "Artificial intelligence neglected these sorts of capabilities for a long time," says MIT AI researcher Tomaso Poggio, "but they turn out to be really difficult to reproduce on machines. After all, evolution has spent millions of years perfecting vision and motor control, but only a few thousand years on language and logic."

Neuroscientists have speculated that the processing of conscious information takes up as little as a thousandth of the human brain's computing power; most of the rest goes into dealing with the lower-level aspects of survival. This suggests that AI has been trying to construct a tower starting from the top; instead of digging right into reasoning and problem solving, perhaps AI should be trying to construct the foundation on which such capabilities rest, as is apparently the case in living creatures.

Poggio and many others have tried to approach the capabilities of vision, motor control, and other low-level functions with a so-called computational approach to AI first championed in the 1970s by David Marr. This approach seeks to analyze low-level brain functions in living organisms and then to distill from this analysis a rigorous mathematical description of the processes that take place; in theory, such mathematical representations can then be implemented on a computer. As Marr wrote: "The mysteries of development and of the central nervous system will ultimately be explained in terms of processes, data structures, virtual machines, algorithms and the particularities of their implementation, control structures, and types and styles of representation of knowledge together with detailed specifications of the knowledge required for different tasks."

Poggio is now one of the best-known practitioners of the computational approach to vision. Having started off by observing the behavior of flies tied to strings, he now heads the MIT AI Lab's "Vision Machine" project, a longstanding, multimillion-dollar effort to reproduce the basic functions of biological vision on a supercomputer the size of a minivan. So far, he says, the machine sees about as well as a fly.

Downstairs from Poggio, in the basement of the AI Lab, Marc Raibert is attempting to do much the same with legged robots. Though legged robots are in general much harder to program than wheeled robots, Raibert notes that nature's choice of legs over wheels is not without merit: 70 percent of the Earth's above-water surface is inaccessible to wheels. For that reason, roboticists have become increasingly interested in legged robots. Raibert is the world's foremost specialist in robot legs; he has been studying nothing but robot legs for over a decade. Raibert has built robots that hop on one leg pogostick style; robots that walk, run, and skip on two legs; robots that trot, canter, and gallop on four legs; and even robots that do flips. He studies films of leaping gazelles, climbing mountain lions, and other animals for inspiration, and, to achieve a firsthand perspective, he himself runs the four traffic-choked miles from his home to the lab most mornings. Like Poggio, Raibert converts his observations into mathematical equations, and then writes programs to implement them on his odd but captivating robots. "They aren't intelligent in the conventional sense," he notes of his machines, "but there are other sorts of intelligence besides the type that resides in the brain."

Though the computationalists had taken a big step toward nature, some researchers felt it wasn't big enough. Poggio, Raibert, and their colleagues were still taking the approach of wanting to figure intelligence out, to write down its mechanisms as a series of equations or algorithms. It was essentially the same approach as the AI scientists who were trying to write chess-playing or mathematical theorem-proving programs, except the computationalists were applying the approach to a lower level of intelligence. What's more, their progress seemed as painfully slow.

In a way, the general public was probably ahead of most artificial-intelligence researchers. After all, by the early 1980s moviegoers had been introduced to the genetically engineered, humanoid "replicants" of *Bladerunner*, and to *The Terminator*, with its organic flesh and brain-like chip. By these standards, even *Star Wars'* whining C3PO, with its metallic body and hollow voice, had come to appear outdated; it seemed almost an article of common sense that an artificially intelligent entity would be, well, *natural* in some fundamental way. Most of AI, on the other hand, was still operating under a paradigm closer to Robbie the Robot from 1956's *The Forbidden Planet*. Needless to say, science shouldn't as a rule have to take its cues from Hollywood. Still, it is ironic that few researchers were prepared to even consider the idea that breaking things open in AI might require trying to get at something more basic in the way nature derived intelligence, something that was qualitatively different from the step-by-step, equation-by equation approach.

One of the first to take the plunge was Stewart Wilson. Wilson is an AI researcher at the Roland Institute, located outside of Kendall Square near MIT, in a discreet building whose main entrance is an unmarked loading dock tucked away from the street. The institute is a scientific think tank, whose generously funded members are free to carry on whatever research happens to interest them. It is here that Wilson, a tall man in his forties who carries the concerned and slightly harried look of a doctor on call, plans ways of approaching intelligence along the same routes that nature did.

It was in the early 1980s that Wilson became convinced something was wrong with the field of AI. "AI projects were masterpieces of programming that dealt with various fragments of human intelligence," he says. "Some of them were amazing, but they were too specialized, and there was no way to generalize them. Another problem was they couldn't take raw input from the world around them; they had to sit there waiting for a human to hand them symbols, and then they manipulated the symbols without knowing what they meant. None of these programs could learn from or adapt to the world around them. Even the simplest animals can do these things, but they had been completely ignored by AI."

What are the roots of intelligence in nature? Wilson mulled over the question for months, and then came up with an answer. "I realized there are deep connections between the need to get food and intelligence, the need to mate and intelligence, the need to survive and intelligence," he explains. "The drive to survival is constantly defining the problem for nature; that's what creates the versatility that living creatures have." Wilson became convinced that intelligence couldn't be effectively replicated until it was placed into the context of the survival of simple creatures.

In a sense, Wilson was merely applying a point that had long been accepted among biologists and psychologists: that the best way to understand how something works in a human is first to understand it in a simpler animal. Since AI ultimately endeavored to replicate human intelligence, Wilson decided the first order of business was to replicate animal intelligence. It was an idea that had never held much currency among AI researchers, but Wilson and others were soon to make it an informal first principle of a new, nature-based approach to AI.

Wilson started playing around with ideas for creating computer simulations or even simple robots that would avoid danger, find food, and deal with their environment in animal-like ways. "If we could do that," he says, "then we can slowly move up to human intelligence, and by the time we got there we wouldn't have left out important characteristics like perception and learning." The idea, he found, tapped into a growing dissatisfaction with traditional AI among his young colleagues, many of whom were ready to jump at the chance to look outside the conventional bounds of the field, to rethink the entire endeavor. Wilson soon took to calling his fake creatures "animatons," but the name didn't catch on until 1984 when he woke up one morning and decided to drop the "on." Soon, researchers at labs around the United States and even Europe were buzzing about the "animat" approach to AI.

It was about this time that Rodney Brooks started to think about insects.

Brooks grew up in the Australian city of Adelaide, but he did not have a typical Australian childhood. "The idea of building intelligent machines was just enormously fascinating to me by the time I was ten," he recounts. "There wasn't much technology available where I was, but I got a few books that described weird stuff, and I spent the rest of my youth trying to build robots. All my friends thought I was a total geek. My summers were spent in this tin shed in our backyard in 105-degree heat removing the transistors from old surplus computers one by one and getting a lot of shocks. I sweated for months over problems I could instantly diagnose now. It was great, great fun." Among his most prized constructions: a robotic turtle and an invincible tic-tac-toe machine, the latter of which left his parents feeling uneasy— "They were convinced I was somehow cheating," he says.

Brooks enrolled at the Flinders University of South Australia, where he eventually entered into a PhD program in mathematics. But his heart wasn't really in the equations; he was spending all his spare time grabbing time on the school's computer to write artificial intelligence programs. Eventually, he bailed out of the program with a master's degree and ended up at Stanford's prestigious AI department. After earning a PhD from his work in machine vision, he spent a little time at Carnegie-Mellon University's AI group, did postdoctorate work at MIT's AI Lab, went back to Stanford to join the faculty (using his spare time to found an artificial intelligence software company that now employs sixty-five people), and then went back to Cambridge to become a professor at MIT.

When Brooks started to cast about for new ideas for research projects, he found himself thinking more and more about mobile robots, and how slow and inept their logical control programs were at navigation. "I knew that couldn't be the only way to do it," he says. "Insects have immensely slow computers with just a few hundred thousand neurons, and yet they fly around in real time and avoid stuff. They're doing far fewer computations per second than that robot's computer was. Insects must organize their intelligence in some better way that allows them to get around so well, and that started me thinking about how to reorganize a robot's computations so it could get around in the real world in real time."

The problem with robotics, decided Brooks, was one of scientific method as much as science: his fellow researchers were simply designing solutions to problems that didn't exist, only to end up overwhelmed by the problems that did. What good were complex vision systems and planning programs, he wondered, if they didn't result in a robot that could deal with the unpredictability of everyday environments? In particular, Brooks frowned on the standard robotics practice of spending months and even years perfecting control programs before actually building a robot—a syndrome he refers to as "puzzlitis."

Brooks's goal was to build robots quickly, simply, and cheaply and get them to do as much as possible in the dirty, fast-changing real world. "The traditional approach has been, 'Let's assume the real world is a static place for now, and after we solve all the problems for that world we'll come back and look at the dynamic world,'" he says. "I just turned that around and decided we should assume the world is dynamic from the very start so we wouldn't get into this trap of doing infinite amounts of computation. I wanted to force the issue right from day one, and insist that robots have got to be able to operate in a world that is changing out from under you."

It was then that Brooks got the idea of focusing on the robot's behaviors, rather than on its ability to process information. "Instead of building the ultimate vision system and the ultimate planning system and the ultimate execution monitor," he says, "I decided to build a robot that could go down a corridor without hitting stuff, even when people were getting in its way." That approach led to some surprisingly simple solutions to problems that had been tripping up researchers for years. Instead of having to design separate programs for helping the robot avoid walls and avoid moving people, for example, Brooks discovered he could lump the two into the simple behavior he calls "avoiding stuff." That, in turn, made it easier to build a vision system. "People had thought that robot vision meant reconstructing a model of the world to know where and what things are," he explains. "But when you view the goal as simply avoiding stuff, then you realize that what you want to know from vision is the simpler problem of where stuff isn't."

The resulting architecture designed by Brooks eschews the abstract reasoning and mapmaking of conventional robot navigation programs. Instead, the subsumption architecture simply lays out a series of straightforward "behaviors" for the robots, such as "wandering around," "avoiding things," and "backing up." Attila has no central brain for deciding which of these behaviors to engage in at any particular time. Instead, all the behaviors act as independent elements that compete for control of the robot; the winner is determined by what the robot sees and feels in the world around it. If one of Attila's light sensors detects an object moving into its field of vision, for example, the sensor might trigger the "avoiding things" behavior, at which point all other behaviors would be temporarily "subsumed." In effect, Brooks is taking exactly the opposite approach from the one traditional AI has been pushing for forty years. He wants his systems to *know* less, to be devoid of even the rudiments of reasoning; in Brooks's scheme, reflex is everything, knowledge and planning are distractions.

A robot under the control of the subsumption architecture, contends Brooks, is capable of far more than the sum of its individual behaviors. Traditional robots are restricted to preprogrammed behaviors, behaviors that can prove unsuitable when faced with a situation unanticipated by its designer. Attila, on the other hand, can ad-lib and, in a sense, invent: even though the rules for firing a particular behavior are straightforward, the ever-changing way in which the rush of sensory information from the real world invokes these rules produces "emergent" behaviors—that is, the simpler behaviors end up working together in unpredictable and surprisingly complex ways, much as a fighter pilot can reflexively fashion intricate maneuvers out of a repertoire of simple actions. In trying to negotiate an odd-shaped obstacle, for example, Attila can artfully invent the appropriate sequence of leg motions for finding footholds and maintaining its balance.

Take the way Attila deals with a brick placed in its path. As one of Attila's up-front "whiskers" hits the brick, it triggers the "lift front leg" behavior. If the front leg doesn't quite clear the brick, a spring-based sensor in the leg "feels" the collision, and triggers the "move leg higher" behavior. If the leg makes it onto the brick on the second try, then the "move forward" behavior is triggered, and Attila starts to pull itself onto the brick. As it does so, the leg sensors start to feel the shift in weight, which triggers the "rotate legs" behavior; all of Attila's legs swing down until they are perpendicular to the ground, ensuring that Attila won't topple over backward as it climbs onto the brick. A similar series of behaviors brings Attila down safely off on the other side.

Though the sequence of movements taken as a whole may be complex and effective, Attila was never explicitly programmed to climb over a brick or any other obstacle; the capability "bubbles up" from the stew of simple behaviors on its own. The idea of intelligent behavior—even if extremely low-level intelligent behavior—emerging rather than having been specifically inserted by a human programmer was a relatively new one to AI, but the concept of emergent properties had been gaining attention throughout the 1980s in a number of research departments, most notably at the Santa Fe Institute in New Mexico. Emergent properties were turning out to be critical to understanding the behaviors of complex systems that had to adapt to change. In particular, emergent properties seemed key to nature's design scheme; the ultimate example was DNA, a molecule that encodes the plans for creating an entire living organism, a model of complexity emerging from simplicity. This elegant concept became the second principle of the nature-based approach to AI: intelligence should be emergent, a property of the complex interaction of simpler elements.

One researcher drawn to Brooks's philosophy was Pattie Maes, a Belgian researcher specializing in designing computer programs capable of learning from their mistakes. Brooks attended a talk Maes gave at a conference a few years ago, and discussed his own work with her afterward. A few months later, Maes was a visiting professor at the MIT AI Lab. Among other things, Maes has written an addition to Attila's control program that allows the robot to teach itself to walk— an impressive feat for a system with no brain. When Attila is first turned on, its legs flail uselessly; each leg is controlled by a completely independent set of behaviors with no clue as to how to coordinate itself with those of the other legs. But as each leg behavior fires, it is provided "positive feedback" and "negative feedback" by, respectively, a small wheel trailing behind Attila that measures forward motion and a button on the robot's belly that gets pushed when Attila collapses. After a little more than a minute, each leg behavior accumulates enough feedback data to figure out how to wait its turn, and Attila is soon cruising along the same way many insects do: the two outer legs on one side move in step with the middle leg on the other side. "It just remembers what works well and what doesn't, and then it does the things that work," explains Maes.

Maes is also working on a method of programming in "motivations," such as "aggression," "curiosity," and "hunger," that would influence a robot's choice of behaviors, as well as a scheme that would allow two robots to achieve a rudimentary form of communication by observing each other's behaviors and adjusting their own behaviors accordingly. "Our critics say that robots without a central model of the world couldn't possibly communicate or demonstrate other advanced behaviors," says Maes. "We're trying to find out how far we can go with our approach. It's our obligation as scientists to study these limits."

Though Attila's competing behaviors are strictly distributed from a software point of view—each behavior essentially has its own control program—the hardware architecture is more of a compromise. Brooks tried giving each behavior its own microprocessor in an earlier robot, but that approach turned out to be a waste of computational horsepower, not to mention an added weight burden. Thus while each of Attila's legs do indeed enjoy a dedicated computer to control lower-level functions such as lifting a leg, higher-level behaviors like wandering share the computing resources of four microprocessors. Sensor data are shared, too; signals from the whiskers, for example, are sent to all the microprocessors so each behavior can tell what's going on. But such hardware consolidation is irrelevant to the distributed nature of the software; even a single processor can support thousands of independent, concurrently running programs, providing a form of "virtual distribution."

One of the payoffs to the subsumption architecture's simplicity is that it doesn't take a lot of computing power to run it. Traditional control programs' step-by-step approach of integrating all their sensory data, reconciling the data with abstract models, and then employing the model to pick out an appropriate action, typically requires dozens of chips or even full-sized computers to handle the information-processing burden. By cutting out the middleman and wiring the choice of action directly to the sensory information, Brooks's scheme results in simpler, more efficient programs that can run on a few chips. Software efficiency is often downplayed by AI researchers as a trivial problem that can be addressed later through more compact hardware or software fine-tuning. But the fact is that many AI programs run so slowly that no foreseeable tuning or hardware performance improvement is likely to allow them to handle practical tasks in acceptable time-frames.

Brooks concedes that his robots don't yet do much under the nascent subsumption architecture. But he insists that architecture's modular nature will allow tacking on new, simple behaviors that will in turn lead to the emergence of ever more complex behaviors, enabling his robots to perform more functions in a wider variety of conditions. That's in sharp contrast to conventional robotics, where behaviors are carefully orchestrated within a central control program. Adding functionality to such an arrangement typically requires reorchestrating all the behaviors so they can be efficiently managed by the central control program, causing programming complexities to exponentially explode. No one can accuse Brooks and his group of getting bogged down: the team has churned out some fifteen different robots in the past seven years, including versions with three wheels, four wheels, and tractor treads, in addition to a number of legged robots. In fact, Brooks's mechanical creatures have become the robots of choice for many researchers working on control programs—so much so that he has set up a company to market them.

And even if his robots are a little dumber than others', says Brooks, they'll still get the job done sooner. "Everyone else gets tied up with trying to deal with theoretical problems that don't arise very often in the real world," he says. "I get very frustrated when people say to me, 'Yeah, but your robots don't do such-and-such.' Well *of course* they don't. Chess-playing programs don't climb mountains, either."

Brooks's ideas haven't exactly unified the field; most AI researchers aren't ready to jettison all aspects of central decision making and maps, aspects that Brooks loudly decries as a waste of time. "I'm glad people have finally realized that there needs to be a reactive component to their systems," he says. "But my point is you don't need the traditional component, and people are finding that a lot harder to accept."

To say the least. Brooks's hubris has goaded AI researchers pursuing more traditional inquiry to fire back, including—or perhaps especially—his associates at MIT.

AI dignitary and MIT AI Lab cofounder Marvin Minsky acknowledges the importance of low-level artificial intelligence, but finds it maddening that the lab, along with most robotics groups, has focused on it to the exclusion of efforts to build intelligent machines. "I love robots, but the people working on robotics there now think the important thing is to make motors and gears that work well in the real world," he says. "They're full of shit. You want a robot that understands what you say and has common sense. These people are making the students waste years having things slip on the floor. It's probably good that somebody is doing these things, but I wish they'd do it somewhere else." Minsky believes that Brooks is dooming his robots to near-uselessness by refusing to imbue them with conventional AI programs' ability to deal with abstract concepts such as time, or physical entities. "Why bother building a robot that's capable of getting from here to there," he says, "if once it gets there it can't tell the difference between a table and a cup of coffee? Or hey, maybe we should all just devote ourselves to replicating insect intelligence."

Poggio is skeptical, too. "A lot of what Rod says makes sense," he concedes. "But a lot of his work is trivial, a throwback to the 1950s. Reflexive behavior can keep a robot from crashing into a wall, but you need a higher level of intelligence to decide whether to turn left or right when you come to an intersection."

The Brooks-inspired split in the MIT AI Lab isn't likely to be resolved soon—nor does everyone want it to be. "I think it's great that everyone is fighting and disagreeing," says lab director Patrick Winston. "They're making things very interesting again, just like it was in the early days." Tomas Lozano-Perez, the lab's associate director, agrees. "Complete agreement is a sign of rigor mortis," he says.

Brooks himself seems to thrive on the criticism. Taped to the outside of his office door for all to see are private evaluations of his work written by leading AI scientists on behalf of various scientific journals and funding bodies. "'This paper is an extended wandering complaint that the world does not view the author's work as the salvation of mankind. . . . The author has little understanding of analytical methods and scientific investigation," reads one review of a Brooks academic article. Says another: "The descent from human-level intelligence to 'artificial orthoptera' is not to be recommended. No doubt these mechanical insects amuse the MIT graduate students."

But though he's gotten the cold shoulder from much of the AI field, Brooks has also been winning his share of admirers; he's shaken up the field, and caused many researchers to reassess their approaches. "The subsumption architecture is wonderful, the best real-time system out there at this time," says Ronald Arkin, a robotics scientist at the Georgia Institute of Technology who incorporates behavior-based controls into George, his own lab's mobile robot. David Payton, now at the Hughes Research Labs in Malibu, California, has adapted a version of the subsumption architecture into his latest robot vehicles. "One of the interesting things we found out," says Payton, "is that in some ways our use of maps was probably inappropriate." Even Red Whittaker says that a Brooks-like element could be the next major improvement to Dante, Ambler, and other robots in his lab.

Brooks, meanwhile, intends to apply his subsumption architecture to a broader array of problems. He has recently been working on a scheme that will allow twenty robots built around toy bulldozers to work together to assemble simple structures—even though no one robot has any idea of what the group is trying to accomplish. It may sound futile, but Brooks points out there's plenty of precedent in nature. "A bee or ant doesn't have a sort of global plan about what's going on," he says. "They're individually following some dumb rules, and out of that emerges a very complex structure like a beehive."

Brooks thinks his ideas will change not only robotics but the world. Attila's planned descendants, now in an early stage of design by Brooks's group, will take the form of "gnat robots"—one-millimeter-wide semi-intelligent mechanisms carved out of a single crumb of silicon, brains, motor, and all, at an estimated eventual cost of pennies apiece. Gnat robots equipped with tiny scalpels, Brooks says, may crawl onto your eyeball or into your heart arteries to perform surgery. They will live in your carpet, continually carrying off dirt speck by speck. Swarms of them will cover your house instead of a coat of paint, flipping over on command when you feel like a change of scenery. "This is for real," he says. "We think we're going to make these little, tiny things go, maybe by the end of the decade."

That scheme may not be as wacky as it seems, thanks to advances in efforts to fashion micromechanical devices out of solid state materials. Boston University researcher Johannes Smits, for example, has developed a 200-step chip-etching process that neatly produces zinc-oxide slivers thinner than a human hair that serve as gearless, jointless "legs" for a chip-robot. The zinc-oxide legs are "piezoelectric," causing them to bend up to 90 degrees in response to an electric jolt as their constituent molecules twist around like compass needles to line up with the electric field. Smits's entire robot design, including microprocessor brain, solar power cell, legs, and a microphone that would allow sending control signals to the robot's brain via blasts of sound, could be built onto the chip in one process.

If technology really does allow for the production of gnat robots, what Brooks would most like to do with them is load them up with his subsumption architecture and send them off to other planets, where they could explore by surfing the wind or hopping like fleas. In the meantime, though, he'd settle for sending Attila to the moon.

"We think it's possible to do a lunar mission for $12 million in total costs using a private launch company," he says. "For that money we could get two of these things to the moon. They don't even need a ground crew, because they're totally autonomous. We think we could carry a small scientific payload, but just to get them there with camera imaging would be interesting. We could get them into the mountainous regions where there haven't been any landers. Maybe we could even get them into the polar regions, where some people say there may be ice in the permanently shadowed areas. That would be an immensely interesting discovery, since the estimated cost for water on a manned lunar base is $150,000 per gallon. Actually, what we'd really like to do is send stuff to Mars. Wouldn't that be fun?"

Probably. More important, Brooks's work has helped lay the groundwork for an approach to artificial intelligence that differs in concept and implementation from the one that has dominated the field for three decades. That's not to say, though, that AI's ruling class intends to go down quietly.

2. FOUNDERING FATHERS

*To succeed, artificial intelligence needs 1.7
Einsteins, two Maxwells, five Faradays,
and .3 Manhattan Projects.*
—JOHN McCARTHY

Doug Lenat is talking to a visitor when an electronic shriek erupts from one of the two oversized computer terminals dominating his deskscape. "Uh-oh," he says, spinning around in his chair with a practiced push from his black Reeboks that lands him squarely in alignment with the beckoning workstation. A glance at the screen apprises him of the cause for the summons, and he attacks the keyboard with the blurred-keyclick speed that programmers acquire from being impatient to get instructions into the computer. Staring at the screen, waiting for the response, Lenat seems to have retreated from the physical world; it is hard to imagine his brain could be more in sync with that of the machine in front of him if the two were hardwired together. Suddenly, he relaxes. "OK, we're in business," he says to no one in particular before turning back to resume his conversation.

If being on such intimate terms with a computer can at times make the chubby, excitable Lenat appear the quintessential computer nerd, he can at least claim he has picked an unusually stimulating partner. Lenat's system, known as Cyc (as in en*cyc*lopedia), represents the most ambitious effort ever to create an intelligent machine—a machine that, claims Lenat, will know everything the average adult knows, from the name of the first president of the United States to how to change a tire, and that will be capable of reasoning with this knowledge in a plausibly human-like way.

Cyc will change the way we live and work, says Lenat, who is in his midforties. Colleges will use Cyc to provide one-on-one tutoring to students, and stores will keep Cyc on hand to custom-design products for individual consumers. Cyc will make scientific discoveries, apply justice, counsel unhappy couples, and even lurk inside television sets in the form of a chip that will note its owners' viewing tastes and edit shows accordingly. "Passive entertainment will become obsolete," Lenat asserts. "Everything will be customized to your preferred level of sex and violence." As Cyc pours through libraries devouring books to expand its knowledge base, he predicts, pieces of it will be stored in computers around the world, its contents made available through phone lines and radio waves. Cyc's intelligence will flow like electricity through a gigantic, ubiquitous knowledge grid.

If nothing else, Cyc is the boldest test yet of one of artificial intelligence's most cherished assumptions: that an ordinary computer can be hand-primed not merely with data and number-crunching abilities, but with knowledge and reason. It is an assumption that has long driven AI, but in recent years has also come to haunt it. If Lenat succeeds in building a machine that at least brushes up against the foundations of human intelligence, he will have earned himself a comfortable berth in the history of science, and even in human achievement; but if he fails, as most AI researchers now believe he will, Lenat will have become no more than the heir to the mistaken assumptions of conventional AI.

In a sense, Cyc is a contemporary rebirth of a four-decade-old vision. In the years after the semiofficial birth of the field of AI, a milestone generally associated with a small conference held at Dartmouth in the summer of 1956, AI luminaries such as Marvin Minsky, John McCarthy, and Edward Feigenbaum pointed with confidence to the coming era of machines that think. Back at the beginning of the quest, AI's leaders thought that day was as close as twenty years away. Now that assessment appears painfully naive. Randall Davis, associate director of the MIT AI Lab cofounded by Minsky, recalls the bravado with which everyone in the field originally regarded the endeavor: "It was as if the first people to arrive in mainland North America looked around and said, 'This looks like a pretty neat place; it'll probably take ten or fifteen years to fairly populate it.' We didn't know how big the territory was."

How badly did AI's founding fathers underestimate their task? In 1950, mathematician and computer scientist Alan Turing proposed AI's most famous test: when a machine could carry on a convincingly human-like conversation with a person, it would be intelligent. Turing felt sure a computer would pass the test by the year 2000, and for a while most AI researchers thought that time-frame was conservative. In 1991, six leading conversation-simulating programs participated in a widely publicized Turing Test competition held in Boston. (In fact, it was a watered-down version of the test, in that the conversations were restricted to topics of the programmers' choosing.) Here is a typical excerpt from the conversation of the program judged to be the most "human-like" of the group (most of the typos were randomly generated by the computer program to imitate human typing errors):

JUDGE: What is the definition of whimsical conversation?

COMPUTER: Didn't you say well ayway I would imagine whimsical conversation is like a series of knight's moves in chess—skrt of zig-zag rather than following any nparticular line.?Why do you ask? Was that a clever comment, or what? Are such questions much on your mind?

JUDGE: I like the image of knights moving haphazardly across the chess board, does tha t mean there is no thought to whimsical conversation?

COMPUTER: I could swear you just said how do you define whimsical?? How do yku define whimssical? Were you always so sick sick sick? Why do you need to tell me you like th eimage of knights moving haphazardly across the chess board?

JUDGE: I did ask you to define whimsical and I don't know if you gave me a straight answer or not. I usually use the word whimsical to describe a piece of art or maybe a story, but hardly ever to describe conversation.

COMPUTER: I think you said how do yo uknow that I am not a dog.?You seem a little hesitant My god please help me! Do you think it's likely that I gave you a straight answer or not?

JUDGE: Probably I can expect no straight answers from you; can you give me a synonym for whimsical conversation?

COMPUTER: I'm sure you said what is the definition of whimsical conversation??have you tried? Ipso facto nonsensico . . . . .Perhaps in your fantasy we can expect no straight answers from you; can each other.

JUDGE: o.k. do you use whimsical conversation all the time or just at certain times?

COMPUTER: I could swear you just said I like the image of k{nights moving haphazardly across the chess board, does that meanthere is no thought to whimsical conversation? Well anyway I would image whimsical conversation is like a series of knight's moves in chess— sort of zig-zag rather than fo. We were discussing you, not me. How about it?

It is a tribute either to the overwhelming sense of good sportsmanship of the contest judges—or else to their stunning density—that the program responsible for this gibberish was deemed by some of them to be more human-like than some of the humans responding to the same questions. The same program, producing much the same sort of nonsense, won the contest the following year as well, and no promising challengers are yet in sight.

Perhaps an even more telling index of AI's progress is provided by the state of chess-playing computers. Chess seems a natural for AI: compared to something as vague and unbounded as conversation, the rules and goals of chess are clear-cut and the moves relatively constrained. In 1954, computer scientist Allen Newell confidently declared that a chess program would reach Grandmaster status within ten years. In fact, it took thirty-four years, when a computer called

"Deep Thought" beat Grandmaster Brent Larsen in 1988. But more to the point, the developers of chess-playing programs had long since shied away from efforts to build into their software a deep understanding of chess strategy and a human-like ability to zero in on the few most promising patterns of moves. Instead, the best chess-playing programs now take advantage of improvements in computer speed to quickly churn through millions of possible moves with calculator-like efficiency in an attempt to find a sequence of moves that most closely matches one of the library of winning games that are programmed into them. Though a chess-playing program may very well defeat a world champion within the next few years, that would be more a victory for those who design fast computer chips than for those who would build a computer that thinks.

Indeed, the endeavor that might be termed "conventional" AI— that is, the effort to develop a computer system that reasons in a highly ordered, step-by-step fashion—has come up short in virtually every arena in which bold predictions had been made for it, including recognizing objects, controlling robots, discovering mathematical theories, understanding stories, comprehending speech, and many other aspects of machine intelligence. After nearly forty years, there are no true breakthroughs to point at. Where did AI go wrong?

The host at the Wursthaus restaurant eyes John McCarthy warily. Despite heavy cosmetic architectural surgery in the 1980s, Harvard Square still retains a portion of the corps of homeless and semidemented denizens that gravitated to the square in the drug-sodden sixties. In his shabby raincoat, sporting a wild, white beard and wired-together glasses, and mumbling in a vaguely belligerent fashion, McCarthy merits a careful screening. He gets his table, but it proves a pyrrhic victory, as his sensitive stomach is almost instantly devastated by his first bite of knockwurst. Between spasms, McCarthy dutifully, if somewhat testily, discusses the work that has consumed him for forty years: boiling all forms of human reasoning down to a system of equations that can be manipulated by a computer. "There is no reason to believe," he says, "that we won't be able to write out the rules that would let a computer think the way we do."

McCarthy cofounded the MIT AI Lab with Minsky in 1958, before splitting for the opposite coast in 1962 to start up AI research at Stanford, where he is still a professor. Recently he was a visiting professor at Harvard, but physical location is often only tangentially relevant to McCarthy's existence; much of his interaction with humans occurs via electronic mail, which he spends hours a week reading and writing in a dimly lit room, bathed in the cathode ray glow of one of the big-screen, high-resolution, black-and-white terminals that are standard equipment for AI researchers.

McCarthy was one of the first, and remains the best-known, proponents of the "logic" school of AI. This camp holds that thinking can be accomplished through the formal language of "first-order predicate calculus," a strange hybrid of mathematics and English. A typical statement in first-order predicate calculus reads like this:

$$\forall x: \text{Spotted}(x) \supset \text{Dog}(x),$$

which translates to: All things that are spotted are dogs. (But not: All dogs are spotted.)

Reasoning with first-order predicate calculus is a matter of pulling together all statements that are relevant to a situation, and then sifting through them in an attempt to draw a conclusion. For example, suppose one were armed with the following statements, or "beliefs," as logicians sometimes call them:

> All dogs have four legs.
> All four-legged animals live on land.
> All land animals have lungs.
> Champ is a dog.

If these statements are true, they easily allow one to answer such questions as, Does Champ have lungs? Do dogs live on land or in water? and Are there any four-legged animals that don't have lungs? Such statements and questions can easily be converted into a computer language, allowing machines to make "inferences"—that is, to answer the questions based on information in the statements. McCarthy and many other researchers have programmed computers to reason reliably in this way, but only under highly restricted conditions. Not surprisingly, the world at large—and especially the quirky world of humans—simply doesn't readily translate into the language of logic.

AI logicians face a raft of problems in attempting to tidy up and formalize the messy, sprawling drama of everyday thought. For starters, even the most useful statements about a situation are likely to prove false at least once in a while. For example, the statement, "Dogs have four legs," seems reasonable, but every so often one encounters a dog with three legs. Furthermore, real-world knowledge typically includes uncertainty: "Rain is likely tonight," or "The taxi will probably be here within ten minutes." In some cases, a perfectly logical conclusion turns out to be invalid in the face of new information: given the statements "It is raining" and "George has an umbrella," it seems fair to conclude that George can take a walk outside without getting wet— until one finds out that George's umbrella is badly ripped. In addition, things change; a system that is told that "Fred has a bus driver's license" will reason incorrectly about Fred's qualifications as a bus driver if Fred loses his license. Finally, people sometimes move outside the bounds of logic, to impressive effect. Lincoln's decision to abolish slavery, Watson and Crick's discovery of DNA's double-helical shape, the surprising curves of a Frank Lloyd Wright home—the intuitions, convictions, and creativity that lead to such accomplishments, and even to far more ordinary ones—typically rely on what a logician would refer to as "unsound" reasoning.

While such issues are not taxing to our freewheeling style of thought, they pose deep challenges to the rigid constraints of the logical approach to AI. McCarthy, for example, has spent much of the last decade experimenting with different ways to bend the rules of first-order predicate calculus in an effort to enable it to deal with the problem of "nonmonotonic knowledge"—that is, situations in which new information can render a previous conclusion invalid. "Logic isn't incompatible with this kind of knowledge," he asserts. "It just isn't obvious how to implement the necessary modifications."

Another keeper of the logical faith in AI is fellow Stanford researcher Nils Nilsson, a tall, vigorous man in his fifties with incongruously white hair. In the mellifluous tones of a radio announcer, Nilsson enthusiastically defends the logical approach against those who would dismiss its shortcomings as fatal flaws. "There are many ways to work uncertain or changing knowledge into first-order predicate calculus,"

he insists, "even if none of these ways have yet proven fully workable."
What's more, logical AI's difficulties are actually something of a virtue,
he adds. As he wrote in an AI research journal article entitled "Logic
and Artificial Intelligence":

> It is important to stress that these conceptualization prob-
> lems do not arise simply as an undesirable side effect of the use
> of logic. They must be confronted and resolved by any approach
> that attempts to represent knowledge of the world by sentence-
> like, declarative structures. The fact that these problems are ex-
> posed quite clearly in the coherent framework provided by the
> logical approach should be counted as an advantage.

In other words, the same problems will turn up sooner or later in *every*
approach to AI—it's just that logic forces them into the open where
researchers can have a clear shot at them.

Most AI researchers aren't buying it. As AI scientist Lawrence Birn-
baum responded in the same journal, in an article entitled "Rigor Mortis: A Response to Nilsson's 'Logic and Artificial Intelligence'":

> I find it difficult to understand the zeal with which logicists
> embrace and defend a theory that has so many problematic im-
> plications. Trying to define "knowledge" and "belief" at our cur-
> rent stage in theorizing about the mind is like biologists trying
> to define "life" a hundred years ago. Rather than seeing this as
> a complicated puzzle to be resolved by artificial intelligence and
> other cognitive sciences as they progress, logicists assume that
> the question has a simple, definitive answer, that logic has pro-
> vided this definitive answer, and that all AI has to do is work out
> the details.

In fact, McCarthy's and Nilsson's efforts notwithstanding, the purely
logical approach to AI has been steadily falling out of favor with the
AI community over the past twenty years.

Roger Schank's computer system could tell you a story. In fact, it could
tell you more than a thousand stories. Schank's story-telling system
comes in several versions, including one that helps children learn

about animals, another that assists executives in business decision making, and a third that aids military officers in logistics planning by relating two-minute videos of any of several thousand snippets of interviews with colonels and generals involved in the Gulf War. In each case, the system accepts some form of "query," or question, from the user, and then sifts through indexes to determine which few stories would be of most relevance. Then the system offers up additional stories it deems to be of related interest, forging a chain of text boxes, talking heads, nature scenes, and whatever else provides worthwhile information.

Schank maintains that his systems are a form of AI he calls "case-based reasoning," in which the chain of "cases," or stories, simulates the sorts of interchanges that enable humans to share knowledge. Useful, but is it really artificial intelligence? Schank bristles at the question. "People are always asking about intelligent machines," he snarls. "The equalling of human intelligence is a writer's and a philosopher's question. Why do we need a machine to think like a human? We have people to do that. What's important is whether or not I can enjoy talking to a machine and can learn from it." Besides, he adds, if there ever is going to be a machine that learns and thinks, it will probably be built on a system like his. "I think this is an intermediate step before the Great Intelligent Machine," he says. "Intelligence is understanding past experiences and reasoning from them."

Schank's view is controversial, not only because it seems a strange, incomplete interpretation of intelligence, but also because it comes from Schank. The hefty, balding, thickly bearded scientist looks the part of the AI guru, and plays it to the hilt; he is as quick to dispatch other approaches to AI as he is to promote his own. He used to hold sway at Yale's AI department, but Schank now has an even sturdier and more visible platform from which to hold forth these days, having scored something of an employment coup: he got the well-heeled management consulting company Andersen Consulting to back him in founding Northwestern University's Institute for the Learning Sciences, which with 150 researchers and programmers became overnight one of the largest AI departments in the world at its birth in 1989. Andersen hopes to recoup its investment in the form of business-oriented AI systems. Schank makes no apologies for giving his endeav-

ors such a commercial slant. "I have to go where the money is," he
once said. "It's been that way since the fifteenth century. The researcher has to have a patron, be it the Medicis or whoever."

Schank's approach is one incarnation of a general approach to AI
that has long been the favorite of the business world. Known as "expert systems," this approach is based not on the precise, formal world
of logic but rather on the messy but practical concept of "heuristics"—
that is, those "rules of thumb" we normally gain through experience
or training. For example, an expert system designed to diagnose car
trouble might be loaded with fifty or so "if-then" rules such as, "If the
engine won't turn over, then check the battery strength"; and "If the
car won't turn over and the battery is charged, then check the spark
plug and ignition coil cables." In operation, such a system would ask
the user to type in responses to questions such as, "Is the outside temperature below 20 degrees?" After checking to see which of the rules' "if"
statements matched the user's responses, it would return the corresponding "then" statement, which might be a recommended action
or another question. Eventually, the user's responses should lead the
system to a single rule containing the proposed diagnosis, such as,
"Replace the battery cable terminals."

Though it's a relatively simple concept, a top-notch expert system
can analyze highly complex situations by considering hundreds or
even thousands of rules to come up with the same results that a human expert would. When practical expert systems emerged in the
early 1980s, the business world was aglow at the notion that computers could serve as far less costly and perhaps more reliable replacements for loan officers, insurance adjusters, geophysicists, and many
other positions requiring years of experience. By 1985 hundreds of AI
researchers had left their positions in academia to join any of the several dozen expert-system companies that sprang up to meet the anticipated demand for this new technology.

The expert system boom turned out to be a bust, however, as researchers found it harder than expected to express a human expert's
knowledge in the form of simple rules. From the windows of the MIT
Lab alone one can look down and see the former headquarters of the
half-dozen companies that made up "AI Alley," now all out of business. That's not to say expert systems don't work; there are thousands

in use today. But there are far fewer than had been expected, and they tend to address a small range of less complex applications, such as checking warranty claims, or watching over parts of chemical manufacturing processes.

Schank and other researchers taking the heuristics approach to AI hope to extend its usefulness beyond such restricted applications, but there is little talk of making a rule- or case-based reasoning system that thinks like a person. If the logicists have failed in their efforts to give machines the ability to reason in a general, powerful way, then the heuristic proponents have succeeded by lowering their sights.

Marvin Minsky is lecturing to a group of faculty and students at Northeastern University. For someone who has spent most of his life thinking about the mind, he presents an extraordinary physical presence. His large skull, virtually hairless except for eyebrows, floats in sharp contrast to the black turtleneck, black slacks, and black sneakers that constitute his preferred uniform. A rigid smile, which might also be called a polite sneer, is more or less fixed on his face. His stooped posture and awkward movements suggest a certain disarray within that cannot be entirely contained.

On this day, the celebrated founding father of the science of artificial intelligence is leaking disarray at an alarming rate. Pacing around the podium, Minsky occasionally rushes over to a table to search in vain through his now-scattered overhead transparencies, and inadvertently switches off the projector several times with swipes of his arms. He isn't so much lecturing as he is blurting out thoughts that seem to suddenly come to him. "Why does pain hurt?" he asks. "What are emotions? How can we spend large amounts of the gross national product on music, when we don't have the slightest idea why people listen to it? There's more music than cocaine; who's cleaning *that* up?" The standing-room-only audience, initially rapt, has stopped taking notes, and some exchange puzzled glances; two people have fallen asleep. One of Minsky's points, if there can be said to be a point, seems to be that everyone has been asking the wrong questions about intelligence for several thousand years. "It's a conspiracy," he says, for the third time.

Over lunch the next day at a Chinese restaurant around the corner from his Brookline home, Minsky concedes that the talk slipped away from him. "I never prepare speeches," he says with a shrug. "They're hit or miss. That was a miss." Minsky appears to be in a foul mood. Then again, Minsky often appears to be in a foul mood, and for good reason: he cultivates anger, he claims, because it helps him think. "Anger is mental hygiene," he says. "I'm angry at myself. I'm angry at civilization." Right now, most of all, he is angry at the MIT Artificial Intelligence Laboratory, which he founded more than three decades ago, and where he is going after lunch. A Minsky visit to the lab is a relatively rare event. "Why should I spend time there?" he says. "They don't pay attention to my ideas, so I might as well be somewhere else."

Though Minsky still holds a professorship at the lab and occasionally teaches a seminar there, he spends most of his time at his large, comfortably cluttered home, surrounded by two player pianos, twelve computers ("Most of them don't get turned on very often," he explains), and vast piles of papers. What he works on, and what he has been working on for the last fifteen years, is a theory he calls the "Society of Mind," which contends that the only way to build an intelligent computer program is through a confederation of less intelligent programs. Minsky fervently believes the theory explains why the science of artificial intelligence has made surprisingly little progress since he helped create the field thirty-five years ago. The Society of Mind, he insists, is the key to getting over the hump.

The man who started people thinking about thinking machines has been obsessed with the science of thought since his days at the renowned Bronx High School of Science. In the mid-1950s, having accumulated mathematics degrees from both Harvard and Princeton, Minsky directed his curiosity toward building a machine that could learn. After briefly trying out faculty positions at Tufts and Harvard, he settled at MIT in 1958. There he immediately cofounded the AI lab with John McCarthy, whom he had known from undergraduate days at Harvard, and who left a few years later to start up Stanford's AI lab.

Minsky was from the beginning of AI a dominant figure who spun off seminal ideas nearly as fast as the growing AI community could digest them. In the lab's first ten years, Minsky, along with later cochairman Seymour Papert, led the group in pioneering explorations of what have become some of AI's most fertile and challenging territories. He and his students wrote programs that enabled computers, under restricted conditions, to prove mathematical theorems, reason by analogy, and understand written English. In 1969, Minsky and Papert published a book that appeared to reveal the theoretical limitations of a type of AI program called a "perceptron," which was based on a theory of how our brains learn by establishing connections between neurons. Though perceptrons had become one of the most popular branches of AI, the influence of the book was such that the field died almost overnight. (More on this later.)

In those early days, it seemed reasonable to strive for the goal of the Intelligent Machine: computers that could solve a range of thorny intellectual problems and reason broadly with common sense, and robots that could figure out how to accomplish tasks with perceptual and physical dexterity. These aspirations were lent vivid, tangible imagery in Minsky's and others' minds by books such as Isaac Asimov's *I, Robot.* "I've always read science fiction for ideas," Minsky says. "These writers think a lot about what an artificial mind would do, what it would need. They put things together in a plausible way." To Minsky and the group of graduate students that quickly gravitated to him—a group that, even in a school celebrated for its students' eccentricity and almost unseemly dedication to technical achievement, became known as one of the more eccentric and dedicated—there was nothing implausible in the idea of creating an intelligent machine in a decade or two.

In what was supposed to be a straightforward step toward his goal, Minsky in the late 1960s directed the resources of the lab toward building a robot that could look at a tower or house built out of toy blocks and construct a similar one. The project, to Minsky's surprise, did not go well. Its vision system had trouble seeing the blocks; its motor control software couldn't place the blocks precisely; and a lack of common sense led it to try building its first tower from the top down, releasing the blocks in midair. "It got to be a big system," Minsky recounts, "with a dozen young people writing hundreds of programs that were getting bigger and bigger, until it finally got to the point where no one understood it much." By 1971, Minsky had abandoned the project. A short time later, tired of fund-raising and other administrative burdens, he passed the chairmanship of the lab onto a young professor named Patrick Winston and set to mulling over what had happened with the robot project.

He soon felt he had begun to understand where he, and everyone else in AI, had gone wrong. At the root of the problem, he contends, is the fact that AI researchers have blindly adopted the straightforward cause-and-effect paradigm that rules in physics and the other hard sciences, where a phenomenon can usually be understood in terms of a single principle. "But anything you do with your brain probably takes five or ten different mechanisms," he says. "Suppose I want this glass of water in front of me. It's half a meter away, so I know I can reach for it with my arm without having to get up. How do I know how far away it is? People in robotics are always trying to find a single good way of doing that, and I think they're wasting their lives, because I can quickly mention a dozen different ways that a person does it. For example, I know my head is about a foot above the table, and I can see the glass is at a forty-five-degree angle to my eyes, so I know it's a little more than a foot away. I'm using common sense, too, to know that if this glass is above the table it must be on it and not floating above it. Another way I can rangefind is to use the binocular fusion of my two eyes, though thirty percent of all people can't do that. Most people under fifty can autofocus their eyes' lenses and get ranging information that way. Another way is I know a glass of water is fist-sized, and I can see that this glass of water in front of me looks about as big as my fist does, so it must be about as far away. Of course, you might be fooled by a giant German stein.

"Yet when people write programs they only incorporate one way to do each thing. I realized you had to have several ways of doing things, that you couldn't depend on any one of them working. The right thing to do is once you get a program running fairly well you should start writing another one and then another one, and then write manager programs with enough knowledge to know which programs to turn off and on. That's what the managers in our brains do; they make sure we feel bored when we're not getting anywhere with a problem, so we try something else. That's what I call the Society of Mind theory."

Minsky, who has never been a great believer in rushing ideas into

46

print, spent most of the next ten years developing and refining the theory, extending it to every aspect of intelligence, including memory, learning, and emotions, and suggesting some of the ways his ideas might be implemented. He published the results in 1986 in a carefully but oddly constructed book—each page holds a self-contained minichapter that is broadly linked, neuron-style, to those around it—called *The Society of Mind*. Minsky considered it his magnum opus; he hoped it would stop much of the field of AI in its tracks, as he had done with his book on perceptrons, and set it scurrying to develop and link the many different intelligent agents his theory prescribed.

When asked whether Minsky's theory about linking together a hodgepodge of programs might not provide a remedy for AI's difficulties, MIT AI Lab researcher Tomaso Poggio, like many others in the lab when Minsky is mentioned, flashes a smile that communicates affection, tolerance, and only a hint of irritation. "I am very good friends with Marvin," he says, a claim that few in the lab make. "I think his idea is nice and probably even correct. But it's true that from a scientific point of view it has been largely dismissed."

Part of the explanation for why Minsky's ideas have received such little attention may be found in the MIT AI Lab's culture, which is fractious even by the standards of the unruly AI community at large. If Minsky has become a victim of that culture, says Patrick Winston, who is still the AI Lab's director, he can look to himself for the blame. "Minsky was an imposing figure then, as he is now," he says. "By the nature of his personality he encouraged a kind of egalitarian, free-for-all style." Winston goes on to tell the story of a Japanese visitor to the lab (most lab visitors these days are Japanese) who attended a Minsky seminar at which someone took issue with one of Minsky's assertions. "The visitor turned to me and said with horror, 'What faculty member would dare to argue with Marvin Minsky?' I said to him, 'Actually, that's an undergraduate.'"

Few researchers at the MIT lab or elsewhere actually take issue with the Society of Mind theory. They simply claim that Minsky has failed to show clearly what it is they are supposed to do about it. "Marvin's book is somewhat cryptic," says Randall Davis. "It is a vast profusion of ideas, and doesn't give a good sense of how it can all work together. In a sense that's poetically just, because intelligence itself may just be a whole lot of little things, of little hacks, and there may not in fact be one general principle that integrates all the little hacks."

Minsky dismisses such criticism. "I think of my theory as being specific and fairly detailed," he says. "A lot of my disappointment with the lab is that it's full of people who are confused between solving a problem, and keeping it solved by not looking at any other possible solutions. Their bug is this: they're afraid to find out which things their ideas are bad at. That's what cancels these people out."

Minsky is right in pointing out that few researchers have ever undertaken major projects that combine multiple approaches to AI. At a minimum, for example, it would seem to make sense to take the strongest features of each of the two major camps in conventional AI—logic and heuristics—and try to integrate them into a single, best-of-both-worlds system that could reason broadly with practical knowledge. One reason few such projects have ever been undertaken is that AI researchers tend to fall squarely into the heuristics or logic camp; few develop enthusiasm for or expertise in both realms. Second, no one in AI has ever been able to drum up sufficient interest on the part of funding bodies to make possible an ambitious, broad-based project with a goal of building a general-purpose, human-like reasoning machine—an AI Manhattan Project.

So when Doug Lenat decided to give it a shot, he had the field pretty much to himself.

Lenat had nothing to do with computers until his last year at the University of Pennsylvania in 1971—a relatively late start for AI researchers, many of whom were hackers addicted to computers before they were out of junior high school. After four years of college, Lenat was about to pick up undergraduate degrees in both mathematics and physics, along with a master's degree in applied mathematics, when he stumbled into a course in artificial intelligence. Lenat soon withdrew his applications to graduate schools in math and physics.

It wasn't just that Lenat found AI interesting; it also happened to be an undeveloped field, which fit in well with his aspirations. "I had gotten far enough and good enough at math and physics to know that I wouldn't be the world's greatest mathematician or physicist," he explains. "I knew I wouldn't even be the world's greatest *living* mathematician or physicist. But in AI, I could be a discoverer."

Lenat made good on the promise at Stanford's celebrated computer science department, where he pursued a doctorate. "He was driven," recalls AI scientist Cordell Green, who was Lenat's faculty advisor. "There were a lot of hot people there at the time, but Lenat stood out because he was willing to stay up all night to crunch through the problems to make it work." (It is a work habit that Lenat has never lost.) By the time Lenat was in his third year of the program, he had developed a program called AM. AM was a type of "machine learning" program designed to derive new ideas on its own. Given a set of basic principles, such a program experiments with different ways of combining and modifying these principles and tests the results against data to see if any of its constructions hold true. Few machine learning programs of this type had ever worked, and the field had been virtually abandoned.

Lenat armed AM with basic mathematical information such as how to determine if two sets are equal, and how to perform the inverse of a mathematical operation. When he turned AM loose, it proceeded to make some seven hundred "discoveries." AM had noticed, for example, that there was a certain subset of numbers that when multiplied together in different combinations could yield any whole number; thus the program had discovered prime numbers and the unique factorization theorem. AM even uncovered at least two obscure principles that appeared to be genuinely original, including one that describes the relationship between the number of times a prime number appears as a factor of certain large numbers and the size of the prime number. The program earned the twenty-five-year-old Lenat a prestigious AI award and sudden widespread recognition in the field.

After receiving his doctorate and picking up faculty positions at both Stanford and Carnegie-Mellon, Lenat went on to develop still cleverer machine learning programs. But even the best of them kept running out of steam after an initial burst of discoveries. "The learning you get out of these programs is really only what you preengineer into them," he says. "It's essentially like a spring unwinding. The energy in the spring comes from choosing the right starting facts, and it enables the program to learn a little bit. But the energy runs out long before the knowledge you really need to continue kicks in. In

1983 I got this vision. I saw what was limiting the work I was doing in machine learning, and it was the same problem holding back [English] language processing and expert systems. It was a lack of common sense."

Even the best AI reasoning programs are idiot savants. Though they may be crammed with highly specialized knowledge, they know virtually none of the things that even a child knows about the real world: what a chair is, how a person is related to her grandparents, that an object can't be in two places at the same time. According to Lenat, this lack of "consensus knowledge"—knowledge most of us take for granted—deprives reasoning systems of two of the key techniques humans use to think their way out of new situations: turning from specialized to general knowledge, and analogizing to seemingly unrelated situations. An appliance repairman knows that when all else fails, giving a toaster a whack might work, while a doctor might get an idea for fighting an illness by analogizing to a military conflict. "These avenues are closed off to computers," says Lenat. "They don't have general knowledge to fall back on, and they don't know any far-flung situations to analogize to. They just die horribly."

To put it another way, Lenat was asserting that to learn a lot, you have to know a lot to begin with. It is not a radical idea; it is the main point of one of the most popular contemporary books on education, E. D. Hirsch, Jr.'s *Cultural Literacy*. Lenat was merely insisting that what was obviously true for people was probably true for machines. "If a program only knows a little bit about math," he says, "it's going to be hard to teach it about rutabagas and looking for jobs."

In 1984, Lenat received "an offer he couldn't refuse," as he puts it, to attempt to put his new theory into practice. The offer came from the Microelectronics and Computer Technology Corp. (MCC) in Austin, Texas, the research and development consortium funded jointly by Xerox, Digital Equipment Corp., Kodak, Apple Computer Corp., and others. Lenat moved to Austin (though he still teaches part-time at Stanford) and immediately set to work to make the most out of the ten years and the $25 million budget laid out for him by MCC. His goal: to equip a computer with all the general and commonsense knowledge of the average adult. Considering that until now no one has come close to imbuing a computer with the knowledge of the average four-year-old, this was a daunting task indeed.

Karen Pittman is hunched over a copy of *How to Troubleshoot and Repair Any Small Gas Engine*. When she finishes, she will teach some of the concepts to Cyc. The point is not to build up the program's expertise in lawn mowers, but rather to provide it with an understanding of the concept of tools in the most general sense. "Elevators are tools," explains Pittman. "So is poetry. So is emotional blackmail." This may seem a little subtle, especially for a computer program, but Cyc's education also includes world history, biology, airline travel, and table manners.

Pittman and the other members of Lenat's team are in the process of entering some 100 million morsels of general knowledge and commonsense reasoning into the software "knowledge base" named Cyc. As they round out the seventh year of the project, Lenat insists the ten-year project is right on schedule. "We have to put in ten times as much information as I thought we would," he says, "but we're doing it ten times as fast." What's more, claims Lenat, in two years Cyc will understand English well enough to continue its own education by reading books, newspapers, and magazines and discussing them with "tutors." After a few years of that, he asserts, Cyc will be ready to usher in a new era in computing. In theory, Cyc would serve as a platform for other programs, allowing them to draw on Cyc's general knowledge and common sense. "My hope," says Lenat, "is that by 1999 no one would think of buying a computer without Cyc any more than anyone would think about buying a computer now without word processing software."

Despite MCC's blue-chip sponsorship, and the fact that from certain vantage points its building's interior bears an eerie resemblance to that of a prison, Lenat has managed to set up a working environment that might have been lifted intact from any major graduate school AI department. Young people in T-shirts and sneakers (and, in one case, shower clogs) lean against doorways with labels like "Hacker's Hell" and catch each other up on late-night computing feats. At lunch, which is shared by the group every day, arguments about obscure points of theory are spirited, and punctuated with flying napkins and lime wedges.

The air of informality aside, the task that confronts Lenat and his team of about twelve is deceptively difficult. Gathering knowledge isn't the problem. What's hard is finding a way to represent the knowledge so a machine can make use of it. After all, Cyc can't learn about automobiles by going for a ride in the countryside. For now, each piece of knowledge must be carefully spoonfed to Cyc in exactly the right form. Lenat calls it teaching by brain surgery.

The surgery is carried out in a sort of programming language developed by Lenat and his group specifically for its ability to handle the fuzzy concepts and irregularities that characterize the real world. Called CycL (and utilized by CycLists), it consists of two main components. The first of these is based on the concept of "frames." A frame is a collection of facts and rules that tells Cyc what it needs to know about a "unit," or a particular thing or concept. The frame describing the unit "Doug Lenat," for example, tells Cyc that he is an instance of the unit "Person," that he is a professor, that his ideology is Republican, and that among the people he likes is Cyc administrative director Mary Shepherd (in fact, he lives with her, but apparently no one wanted to get into this with Cyc).

A frame looks like a sort of fact sheet on the computer screen. To build a frame about a particular unit, a Cyc "knowledge enterer" must first decide which kinds of information about the unit will be useful to Cyc. These information categories, or "slots," are then typed in as a list. The frame for the unit Texas, for example, might look something like this in its early stages:

Texas

Capital:
Soil Quality:
Topography:
Controlling Country:

Once the slots are laid out, the knowledge enterer must ensure that Cyc knows what they mean. Thus Cyc needs to have a separate frame constructed for "topology" (if it doesn't already exist) and for every other slot category that appears in the frame. Finally, the slots can be filled in with "values." (In the case of Texas, these would be Austin, Rocky/Sandy, Desert-like/Hilly, and USA, respectively.)

Frames provide a powerful and convenient way of enabling Cyc to understand the relationships between different objects and concepts. A unit that is an instance of another unit, for example, automatically "inherits" some of the other unit's attributes. Thus, because the unit "Chewing" is an instance of the unit "Bodily function," Cyc automatically knows that chewing is one of the things that a person or animal is likely to do on a regular basis.

Like most AI programs that have to deal with the real world, Cyc uses "default logic," which means that Cyc can be told to assume that something is usually true, but that there may be exceptions. Rather than having to specify to Cyc that horses, mice, people, and elephants give birth to their young, Cyc can simply be informed that almost all mammals give birth to their young. If Cyc ran across a reference to a duckbilled platypus and learned it was a mammal, it would incorrectly guess that the platypus followed the birth rule; someone would have to tell Cyc that the platypus lays eggs. That may seem a little inexact, but, in fact, people reason this way, too. "If I asked you where your car was, you would answer under the assumption that your car was where you left it," says Lenat. "Your car may have in fact just been stolen, but it's better to get an occasional unpleasant surprise than to exhibit the irrational behavior that would result from always thinking that your car may have just been stolen."

As Cyc's education has progressed, the emphasis in programming technique has gradually shifted from frames to a component of CycL known as "the constraint language." Essentially a form of predicate calculus, the constraint language allows teaching Cyc more complicated concepts. Take, for example, the information that a person named Fred owns either a parakeet or a dog. Expressing this idea in terms of simple slots and values would require defining to Cyc a new category consisting of animals that are either dogs or parakeets. To avoid bogging Cyc down with such otherwise useless categories, the constraint language allows writing something like, "There exists x such that x is either a parakeet or a dog." This statement can then be stuck into the slot for pets in the frame describing Fred.

Like us, Cyc also has to know how to ignore irrelevant information, lest it waste too much time "thinking" or provide answers that are correct but useless. "When we're deciding whether or not to lock our car, we don't ask ourselves how many legs a spider has or what we had for dinner last night," says Lenat. "And if someone asked you if there had recently been a large quantity of blood in your car, you probably wouldn't take into account the fact that your body is a large bag of blood." Cyc has ways of avoiding such impracticalities. It knows, for example, to start its search with the most obviously relevant rules, and then to branch out to more subtly related concepts. Thus if asked a question about freeways, Cyc might start with its knowledge about freeways, then move on to roads, cars, transportation, and so on, presumably coming across its answer long before it got to spiders. Cyc can also be told that in many situations people tend to overlook the fact that they contain blood, allowing Cyc to do the same.

To further improve its speed, Cyc can access any of about two dozen different reasoning techniques, each one of which is best suited for a certain type of problem. For complex problems, Cyc may have to sift through many thousands of rules one by one to arrive at an answer. But if asked how grandparents are related to their grandchildren, Cyc can simply apply inverse reasoning to its knowledge that grandchildren are their grandparents' children's children and conclude that grandparents are their grandchildren's parents' parents. In situations where information is incomplete or uncertain, Cyc can try to fill in the blanks and assess plausibility to come up with a best guess. If Cyc is told that a person has voted and the person's mother is thirty-seven years old, for example, Cyc might guess that the person is eighteen years old, based on its knowledge of the voting age and the age at which a woman is likely to start having children.

Bringing Cyc to this level of sophistication was an excruciatingly slow process. For the first year of the project Lenat and his team hardly went near a keyboard. Why bother trying to teach Cyc what a truck is if it didn't understand cars, wheels, or even motion? The group had to start by defining such fundamental concepts such as space, time, action, and physical objects, and by deciding how to go about organizing complex ideas and things into terms that Cyc could handle. "We sat around for four months with enormous pieces of white paper trying to determine what categories were worth distinguishing in the world," recalls Lenat. "A lot of our insights were surprising, like when we suddenly realized that what we think of as tangible objects, such as a particular person or a table, are actually events with a starting time, an ending time, and a duration. We don't normally like to think of it this way, but people are executing the process of existence."

The team also found it necessary to push Cyc into the Zen-like view that people and objects have physical and nonphysical components that must be regarded as separate entities. If asked to provide the age of a book, for example, Cyc would have to determine whether it was appropriate to consider the date that the physical book itself rolled off the presses, or the date that the content of the book came into being—that is, when the author finished the manuscript.

The group spent most of the first year thrashing out these representations and definitions. Time was explained to Cyc in terms of fifty different types of relationships between events, such as, "Event A starts after the end of event B." Physical objects were broken down into things that are countable, like marbles, and things that are "stuff," like peanut butter. When they were finally through, Cyc had been given a basic version of what AI researchers call an "ontology," or "worldview," through which it could make sense of reality.

The next two years were spent refining the ontology and building CycL. It was 1990 before Lenat's group began to pour in the knowledge. Actually, drip was more like it. At first, each new concept required constructing a frame from scratch, slot by carefully chosen slot, and each new slot required creating yet another frame to explain what that new slot meant, until finally an entire frame could be defined in terms of slots with which Cyc was already familiar. The process was so painstaking that when Cyc was introduced to its first sentence— "Napoleon died in 1821; Wellington was saddened"—it took two months to teach Cyc what it needed to know to understand it. "We had to explain about time, communication, human emotions, and life and death," says Lenat. "It seemed like a never-ending process."

Fortunately for everyone involved, Cyc learns more quickly now, requiring only seconds to digest a new sentence. It has at times contained over two million pieces of knowledge—about half of which are facts, and half rules—but that number can go down as well as up, as Cyc combines rules by making useful generalizations. Two million is a long way from 100 million, but Lenat's not concerned. As he predicted, the more Cyc knows, the more quickly it learns. "Things can now be defined quickly in terms of other things," he says. "If we want to teach Cyc what a cheetah is, we just call up what it knows about lions and tigers, and make a few changes."

Cyc's growth is partly due to Lenat's efforts to find people who are good at analyzing and entering knowledge. Most people simply don't seem to have a knack for it, but Lenat eventually came across Karen Pittman, who is a botanist by training, and Nick Siegel, who is a cultural anthropologist. "Doug wanted people with some cross-cultural experience, who could examine the implicit knowledge in society," explains Siegel. Though there are now some three dozen people who enter information into Cyc on at least a part-time basis, Pittman and Siegel were for a time the only full-time knowledge enterers, and are responsible for much of what Cyc knows. "I guess I'm becoming an ex-botanist," Pittman says with a laugh.

The knowledge-entering routine is a unique one. Every day, Pittman, Siegel, and others read selected articles from any of a wide variety of publications and ask themselves what Cyc would need to know to understand the article. A group favorite: *The World Weekly News*. "What are the things we know," muses Lenat, "that allow us to reject as untrue an article about the discovery of a human skeleton on the moon?" After teaching the required new knowledge to Cyc, team members ask it questions to see if the lesson sunk in. Having been told that someone drove to work in the morning, for example, Cyc might be asked how that person got home that evening.

Cyc can also ask questions of its programmers to clarify points, resolve apparent contradictions, and fill in gaps. The questions come to Cyc as knowledge is being entered into it, but it politely stores them away until someone asks to review them, at which point they are displayed on the computer screen in plain English. One recent inquiry: "Reproductions make sense for pure information, but someone is using it to apply to paintings and statues. Is it OK for something that is at least partially tangible to have reproductions made of it?" One of its first questions came up when it was told that intelligent things tend to like other intelligent things of the same type, and then was later told that Mary Shepherd liked Cyc. "Am I a person?" asked Cyc. "Or is Mary Shepherd a computer program?"

Every so often, Cyc takes a little time off to "meditate." That is, it looks over its knowledge base in search of interesting analogies, or at least what seem like interesting analogies to Cyc. (Cyc used to meditate all night long, but now it's fast enough to work the meditation into its daytime routines.) These analogies are then examined by Lenat and others as a way of weeding out flaws in the knowledge base. A trivial analogy, such as "warmth" to "heat," would suggest that Cyc doesn't realize how similar the two concepts are, or it would have known that they didn't make for an interesting analogy. A baffling analogy, such as "radio" to "soap," would imply that something hasn't been defined clearly. A truly ingenious analogy, on the other hand, would provide hope that Cyc may yet fulfill Lenat's predictions. So far, Cyc doesn't seem to be having much trouble avoiding the trivial. Baffling is more of a problem, but Cyc also manages to come up with its share of the clever. One night's ruminations included the following blend of off-the-wall and intriguing analogies: "dad" to "dictator," "head of state," and "animal found in regions"; "automobile" to "elevator," "escalator," and "consumer electronics"; "owns" to "physically contains," "is aware of," and "goals"; and "profession" to "prominent for," "occupant of," and "spouse."

If Cyc can learn to digest plain English, the rate of growth of its knowledge base will explode. Instead of relying on a handful of people to type in selected crumbs of information, it could tear into everything from Descartes to *People*, tucking away every fact and insight it could glean. (Most magazines are available in computer-readable form these days, and books can be translated into electronic form by "reading machines" that optically scan a page of text.) But getting a computer to understand English, or any "natural" language, is a much-sought-after goal that has eluded computer scientists for decades. The reason Lenat thinks he will succeed where others have failed is predictable: "You can't do natural language understanding without a knowledge base like Cyc's," he says.

This may sound like circular reasoning—Cyc needs to understand English, and understanding English requires Cyc—but Lenat offers some justification for the argument. Until now, natural language projects have focused on understanding grammar and syntax, under the assumption that if a computer knew what types of words it was dealing with, meaning would be relatively easy to determine. In fact, notes Lenat, pulling out the many possible meanings of a word in a typical sentence can be a Herculean task. "What is a 'red conductor?' " he asks. "Is it a piece of metal that's red? A communist musical conductor? Or take the two sentences, 'Mary saw a bicycle in the store window. She wanted it.' How do we know that she wanted the bicycle, and not the store or the window?" Things get even stickier in the real world, where Cyc would have to deal with newspaper headlines like, "British Left Waffles on Falklands."

Cyc's natural language interface is under construction by a group across the hall from Lenat's team. Though the interface is intended to work with other knowledge bases, and Cyc should be able to accept other interfaces, the two projects are closely linked. "We're leveraging our language knowledge and Cyc's general knowledge against one another," says Kevin Knight, a researcher on the interface team. The interface will be able to break down an English sentence into different possible interpretations, and Cyc can then analyze them, selecting the one that best jibes with its common sense. There are some early signs that the intended synergy is paying off: when informed that someone had "read Melville," Cyc guessed the person had read *Moby Dick*, even though it had never been told the meaning of the expression "to read an author."

But Cyc could fail before it ever gets a chance to read. The key danger comes from "divergence"—that is, when different pieces of knowledge conflict with or obscure one another. A certain amount of divergence is inevitable. Different people are bound to inadvertently provide Cyc with different ways of looking at the same thing, such as when Cyc is told by one person that "father" is a biological father and by another that it is a male parent. Lenat insists that at least for now most such inconsistencies can be caught through Cyc's analogy hunts. Even if some slip through, he points out, Cyc's ability to recognize the fallibility of the "truths" in its knowledge base prevent it from crashing.

But Lenat concedes the growth of the knowledge base could outstrip these protection mechanisms, allowing "fatal" divergence to set in. "If there are too many inconsistencies," he shrugs, "the knowledge base will collapse." That means Cyc would start losing its ability to summon up and make use of all its relevant knowledge when solving a problem, resulting in less discerning or even incorrect responses. But Lenat insists Cyc will manage to avoid this and other potential show-stoppers.

Most researchers are far less optimistic, claiming that Lenat has done a better job in raising expectations for Cyc than he is likely to do in building it. "Doug has hyped the project way beyond what I think is possible," says one computer scientist close to Cyc. "A lot of careful scientists think he's way out on a limb. There are too many thorny AI problems that have to be solved before he can achieve the kinds of results he's after." For example, Cyc needs a better way to decide which of its reasoning processes it will use in a particular problem, says this scientist, as well as a more effective technique for dealing with subjective statements such as, "Democracy is the best form of government." Another key problem with Cyc, say critics, is that it is limited to a single ontology—that is, to one way of categorizing knowledge—and to CycL's particular format for representing information.

Perhaps most telling about Cyc's prospects is that most of the AI leaders who epitomize the ideas that inspired Cyc do not profess to hold out much hope of its achieving common sense or any other forms of human intelligence. Minsky grumbles that Cyc doesn't incorporate enough approaches. Schank predicts Cyc will be buried in static facts that can't keep up with the dynamic world around it. McCarthy questions the adequacy of Cyc's logic. And though Lenat has raised his estimate of his chances of success from an initial 10 percent to 60 percent in 1990, and now most recently to 99 percent, it would appear that this growing self-assurance stems not so much from Cyc's progress as from Lenat's having quietly toned down the scope of his goals. Instead of emphasizing the production of intelligence and common sense, his reports now point out the "valuable learning experience" that creating Cyc has provided. And instead of talking up the idea that Cyc will plough through books and magazines unaided, he now somewhat vaguely predicts that Cyc "will be adequate to serve as a platform for natural language understanding."

If Cyc was envisioned as AI's Washington crossing the Delaware, it is turning out to be more of a Custer's Last Stand. What's more, no successor project is waiting in the wings. If Cyc fizzles, as most believe it will, conventional AI will have nothing to point to as evidence that it knows how to approach the big questions of replicating intelligence. As AI researcher Terry Winograd recently said: "AI is no closer to intelligence than the alchemists were to gold."

A black-clad Minsky is once again sitting before an audience, eating peanuts at a table while waiting to be introduced. Next to him, looking like a homeless person who has been led to the table for a hot meal, is McCarthy. Tonight the two are here to debate fundamental questions about AI—such as "What is intelligence?"—in front of a few hundred AI researchers. The two brilliant curmudgeons have at it for an hour, spending almost as much energy on disparaging the panel of questioners as they do on responding to them. "You're asking the wrong question," snaps Minsky to one. "When you've organized your thoughts send me an E-mail message," growls McCarthy to another. The two are like bookends.

The pointless bickering seems to underscore a truth that has already dawned on much of the audience: Minsky's and McCarthy's vision of a machine that could be stuffed with intelligence is receding from AI's grasp faster than even someone as smart, ambitious, and well funded as Doug Lenat can approach it.

In fact, as far back as the mid-1980s, many young researchers entering the field were already turning elsewhere for inspiration. The inspiration, as it turned out, was everywhere they looked: it was nature itself. The leading figures in AI had always assumed they could transcend nature, that they could analyze human intelligence and design its equal from a clean piece of paper, with little regard for the way intelligence evolved and became embodied in living creatures. But other, nature-based, strategies were soon taking shape within the AI community. One was Rodney Brooks's subsumption architecture. Another comprised a revival of an early branch of AI that had been all but buried some twenty-five years before.

3. THE ART OF THOUGHT

If you were playing God, how would you have done this?
—STEPHEN GROSSBERG

t can pick out and recognize a moving object against an obscuring background. It can accurately reach with a multiply jointed arm holding a tool on its first try. It can identify engine parts, poisonous mushrooms, and the symptoms of pneumonia. It can write in cursive script. It can deal with new information, uncertain information, and incomplete information. It teaches itself, has expectations, and focuses its attention on important events.

There are only two entities in the known universe that possess these capabilities: the human brain and ART. ART, a computer-based model of how the brain works, still lags the brain by quite a bit, but Stephen Grossberg and his group at Boston University's Center for Adaptive Systems have been making surprising progress in closing the gap.

ART is eerily brain-like in function. It comes to conclusions not in the linear, step-by-step fashion that characterizes most AI systems, but rather with a "whoomph" of recognition that seems akin to the sensation we experience when we finally pull a half-forgotten name out of our memory. ART learns as the brain does, by studying examples, picking out patterns, and drawing its own conclusions. It has components analogous to those of human brain; when one of these components is disabled, ART's resulting impairment resembles that of people who have had brain lesions in the corresponding brain section. As Grossberg puts it, ART is "a microcosm of cognition." More formally, ART is an artificial neural network.

An artificial neural network is a computer program or hardware device designed to imitate the brain's reliance on a vast array of interconnected cells known as neurons that act like tiny but complex electronic switches. In Rodney Brooks's robots, a set of behavior programs interacts to create new, more complex behaviors. The neural network that is our brain creates behaviors out of the interaction of relatively simple elements, too, but the complexity of the interactions—the sheer combinatorial possibilities of 100 billion neurons constantly chattering among themselves—are so much greater than in those of a Brooksian system as to make the comparison superficial. After all, the results of these neural interactions are not mere obstacle avoidance, but the entirety of human thought.

Given the enormous chasm in complexity between a human brain and the control program of a mobile robot designed to not run into walls, is there any reason to believe that the emergent behavior approach can be "scaled up" to produce systems that even begin to broach the domain of human-like intelligence?

The scaling question has bedeviled AI from the very beginning. Within a few years of AI's debut in the 1950s, scientists had produced programs that could prove mathematical theorems, recognize simple shapes, and understand written stories. These systems were error-prone and severely limited in the types of inputs they could handle. But at the time it seemed their success, limited as it was, proved the validity of the approaches they embodied. Producing more competent systems, most AI researchers agreed, was simply a matter of fine-tuning and broadening the scope of these prototypes. But as with Minsky's tower-building robot, researchers soon found their systems sinking under the growing length and complexity of their programming code. Their approaches simply didn't scale up.

Would a system based on the emergent properties of many simple, interacting elements also necessarily become hopelessly unwieldy as it was made larger and more sophisticated? In fact, there is a reason why such systems might be able to avoid the scaling problem. The escape clause lies with "self-organization," the ability of a system to forge order on its own out of the seemingly chaotic interaction of its elements. All of biology is an example of nature's reliance on self-organization. The fact that life in all its complexity and precision arose out of orderless pools of inorganic matter is a testament to the technique. The structural fingerprints of self-organization are visible on every organism: biological designs that at first glance appear seamless wholes are in fact confederations of smaller structures, which are in turn made up of even smaller alliances, and so forth. Thus we are a system of organs, our organs are composed of valves and membranes and the like, these components are made of cells, cells are constructed of organelles like mitochondria, these elements are made of proteins, and proteins are assembled from amino acids.

Nowhere is there an explicit blueprint for constructing human beings out of amino acids. But simply by obeying the laws of physics and statistics, proteins organize themselves into cells, cells into organs, organs into human beings. It may have taken nature a billion years to fine-tune the process, but now a brand-new human being can construct itself—properly wired brain and all—in nine months. In the same way, we can think, not because we've been loaded with a program spelling out the specifics and nuances of intelligence, but because our neurons follow rules of connection and interaction—rules still largely undiscovered—that cause intelligence to bubble up, layer by self-organizing layer. If an artificial network of neuron-like programs or circuits were given a corresponding set of rules of connection and interaction, then theoretically it should be capable of giving rise to all the forms of thought familiar to us.

Though research in artificial neural networks might logically be considered an extension of the conceptually and practically simpler schemes espoused by Brooks, the first work in this area actually predates Brooks's efforts by three decades. In fact, the foundations of neural network studies were built a century ago when the Spanish biologist Santiago Ramon y Cajal first described the structure of neurons and their astounding interconnectedness. Extending from the neuronal cell, observed Cajal, are a long appendage called the axon and a series of finely branching tentacles called dendrites. A neuron's single axon is typically connected to the dendrites of thousands of other neurons; the point of connection is called a "synapse." When stimulated, a neuron "fires"—that is, an electrical pulse speeds down the axon, where it is picked up by the many dendrites attached to it at synapses and sent on along to those various neurons. If the combined signals picked up from all of a neuron's dendrites are strong enough—signals can be strengthened or weakened at the synapse or at other points along its journey—then that neuron will fire a signal along its own axon, passing the signal on to thousands of other neurons, and so on. In this way a cascade of signals can be promulgated across millions of neurons in the blink of an eye.

Ramon y Cajal surmised that thought and memory arose from these connections between neurons. Even more important, he realized that the brain was constantly rewiring itself, strengthening and weakening various synapses in response to learning and experience; this, apparently, was behind the brain's ability to teach itself new tricks and store new memories. But what, exactly, was the rewiring scheme?

In 1949 novelist-turned-psychologist-turned-neuroscientist Donald Hebb proposed a simple rule for such self-programming: if two connected neurons fire at, or almost at, the same time, then the connection between them gets stronger. The usefulness of such a seemingly artless rule lies with the fact that the more often two neurons happen to fire at the same time, the more likely it is that they're firing together for a good reason—that is, that they are both part of some pattern of neural firings that the brain is employing to represent a remembered word, or part of an image being taken in by the eye, or a newly learned hand movement. Hebb's rule, in other words, provided a simple but effective learning scheme: when neurons repeatedly fire in a particular pattern, that pattern becomes a semipermanent feature of the brain. If the connections so formed should later prove of little use, resulting in infrequent firings across the previously strengthened synapses, the connections weaken on their own from the lack of stimulation. That's why you repeat a phone number to yourself several times to remember it: the repetition is strengthening the neuronal connections that will store the number even after you've stopped repeating it. And that's why the number is likely to fade from memory if you don't recall it for several months, allowing the connections to weaken.

This straightforward scheme for learning requires a shift away from the simplistic notion that information is stored by individual neurons to the more subtle idea that information is stored and expressed in the patterns of firings between many neurons. Individual bits and pieces of memories and thought aren't assigned to particular neurons; there is no neuron, or small patch of neurons, that could be surgically removed so as to remove your memory of your grandmother's face while leaving all other memories and thoughts unimpaired. Instead, the memory of your grandmother's face is embedded in the motif of firings shaped by neuronal connections; and many of these same connections are also employed to create patterns recalling your grandfather, your dog, and your favorite song. That's why we tend to remember things in strings of connections, or associations: when trying to recall the name of someone you met several weeks ago, you may first think of a mutual friend, and then be reminded of an image of a flower, before you recall that the name is Lily.

By the early 1950s, scientists were sufficiently intrigued by the brain's odd but evidently effective information processing scheme to attempt to model it with electronic circuits. These devices typically consisted of a dozen or so "nodes" meant to duplicate, in a rough way, the function of neurons: if a node received a strong enough signal, it would "fire," sending its own signal off to the nodes to which it was attached. These early artificial neural networks had two types of nodes: input nodes, which accepted a signal from the outside world, and output nodes, whose firings represented the network's reaction to a given input. When a pattern of signals, or a "stimulus," was entered into the network via the input nodes, an answering pattern, or "response," was picked up from the output nodes. In this way, these primitive neural networks could classify, or "recognize," certain types of patterns.

Pattern recognition may not at first seem as interesting or important a problem as, say, mathematical theorem proving, but in fact it is central to our existence. The act of looking around a room is an exercise in pattern recognition; a pencil is perceived as a pencil because the information the eye pulls from the pencil provides an input pattern the brain deems familiar. That the sound of conversation has meaning owes to pattern recognition, each syllable being seized on by the brain as matching an established pattern. Much of problem solving is pattern recognition, too: if you try to start your car and the engine doesn't turn over, you don't immediately start to work your way through a step-by-step chain of formal if-then rules. Rather, the brain is capable of instantly recognizing a familiar pattern of conditions: it's cold, the engine isn't turning over, there is plenty of gas in the car—the battery must be low.

Humans are the ultimate pattern recognizers. One-month-old infants almost unfailingly respond to a smiling face, grabbing that image from out of the huge, sprawling jumble of brand-new visual stimulation to which it is exposed. Toddlers are continually mapping the world into useful categories: motorcycles, cars, and trucks are quickly lumped into one category; birds, cats, and dogs in another. As mentioned before, even chessmasters, often popularly but mistakenly regarded as highly structured thinkers who logically churn through alternative moves, are typically master pattern recognizers: the layout of pieces on the board almost instantly triggers insights into promising new patterns.

To understand the basic operations of the simplest type of neural network, imagine three input nodes arranged in a column, and a column of two output nodes to the right of the input nodes. Each of the three input nodes is connected by wires to each of the two output nodes; but in the midpoint of each wire is a switch that can be turned off to prevent electric current from passing through the wire from the input node to the output node. When either type of node is exposed to a large enough current, the node is said to be "active"; if there is too little or no current present, the node is inactive.

To solve a problem with such a network, each input node can be assigned to represent a particular feature of the problem. If the network is to distinguish cats from mice, for example, then the top input node might represent tail type—active for fluffy and inactive for thin; the middle input might represent fur color—active for white and inactive for other colors; and the bottom could represent body length—active for smaller than eight inches and inactive for larger than eight inches. As for the output nodes, a pattern of active/active could arbitrarily be taken to mean mouse, while active/inactive could mean cat. By judiciously setting the switches in the wires, one could easily arrange for an input pattern of "inactive/active/active" (thin tail, white fur, short body) to produce an output pattern of "active/active" (mouse), and for active/inactive/inactive to produce "active/inactive." (The appropriate input pattern could be entered manually; even better, one could hope to hook up the three input nodes to sensors that automatically measured each test animal's tail width, fur color, and body length, respectively.) Other animals could also be added to the network: thin tail, white fur, and long could trigger "inactive/active" to signify a rat, while fluffy tail, nonwhite fur, and short could trigger "inactive/inactive," representing a chipmunk.

The key to having the network produce the right answer, of course, lies with adjusting the switches between the input and output nodes. In the simplest networks, this adjustment may only be a matter of turning each switch either fully off or fully on. But more typically, neural network schemes require that switches be set to any of a range of signal pass-through strengths, so that an input node might contribute all the signal, or only some fraction of the signal, required to fire an output node to which it is connected; in some cases, the switch may even be set to pass on a "negative," or "inhibitory," signal that makes it *less* likely that a connected neuron will fire. In neural network terminology, the settings of these switches are the "weights" of the connections between nodes; increasing the weight between an input node and output node means that the input node signal will make a larger contribution to the likelihood that the connected output node will fire.

Individually calculating the weights required to produce desired pattern matching turned out to be a laborious task on small networks, and virtually impossible on larger networks with exponentially increasing numbers of connections. If artificial neural networks were to be of practical or even conceptual interest, there had to be some sort of shortcut for finding the appropriate weights between input and output nodes. In other words, what was missing was something along the lines of Hebb's rule that could be implemented on an artificial neural network to allow these networks to "learn."

· · ·

The breakthrough came in 1959 from Cornell psychologist Frank Rosenblatt, who designed a neural network called a "perceptron" that could be "trained" to perform simple classifications of visual patterns, based on the input from a few hundred light-sensitive electronic cells. The weight-setting method was simple and Hebb's-rule-like. The perceptron started with its weights set randomly. Then the system "looked" at a pattern—that is, the various input nodes connected to the photocells fired or didn't fire, depending on how much light each photocell detected in a particular region of the image. These input node firings then resulted in some pattern of output node firings, depending on the randomly set weights.

In general, the first answer churned out by the perceptron was wrong, but Rosenblatt trained the network according to the following simple rule: the weights of all the connections leading to output neurons that should have fired but didn't were slightly increased, and the weights of all the connections leading to output neurons that shouldn't have fired but did were slightly decreased. For example, suppose that the desired response to a particular visual pattern included firing the first output node and not the second; and suppose the first time the perceptron looked at the pattern it fired the second but not the first. Training would then have involved increasing the weights of all connections leading to the first output node, increasing the chances that it would fire the next time that pattern was shown, and decreasing the weights of all connections leading to the second output node, decreasing the chances that it would fire the next time that pattern was shown.

After making the required set of adjustments for a particular pattern, the perceptron was shown another pattern, and the weight-adjusting process was repeated. After perhaps a few hundred patterns were presented in this way, the entire training process was repeated on the same set of patterns a second time, a third time, and so on, for as many as thousands of times, until the perceptron came to give correct answers for an acceptably high percentage of the time.

The extensive repetition of the training process was required because the weights were only slightly adjusted after each pattern—so slight that one set of adjustments generally wasn't enough to cause the perceptron to give the right answer the next time. But keeping the adjustments slight was critical to the perceptron's reliability: if the weights were heavily adjusted so as to immediately produce a correct answer for a given pattern, then the radical adjustment would have wiped out all previous training and would have resulted in the wrong answer for all other patterns. By keeping each adjustment slight, the network could gradually edge closer to a single, "stable" setting that, while not perfectly suited to identifying a single input pattern, was sufficiently suited to identifying most patterns. In fact, a well-trained perceptron was capable of classifying not only those patterns on which it had been trained, but also new patterns of the same type— for example, letters of the alphabet printed in slightly different fonts.

In essence, the perceptron training process was a form of programming, and the resulting weight settings were a type of program. But it was a very different sort of program than the traditional step-by-step program familiar to computer scientists. The perceptron was, in a sense, writing its own program based on a simple rule and on information provided by the trainer as to which output nodes were giving right answers and which wrong answers. Clearly, a perceptron program—that is, the weight settings—embodied some sort of logic, a set of pattern-classifying rules. But it was not a set of rules that could be easily analyzed or distilled into conventional logic statements; the intelligence embedded in the program was opaque to observers. But it was a rudimentary form of intelligence nonetheless.

The perceptron represented an entirely new approach to AI. Instead of attempting to duplicate the brain's achievements in high-level thought, the perceptron was an attempt to simulate the brain's *style* of computing at a fairly low level—recognizing simple patterns instead of trying to prove mathematical theorems. The idea was to build intelligence, not from deep principles of logic or complex algorithms, but rather at the level of simple switches; the more complex modes of thought, it was hoped, would eventually rise up out of these foundations. As one scientist recently put it, the motto of neural network researchers could be, "Build it and it will come"—that is, if the right switches are given the right rules, intelligence will emerge. Though it may seem something of a shot in the dark, the neural network approach is supported by one intriguing piece of evidence: our brains seem to work that way.

In fact, though Rosenblatt and other neural network pioneers didn't articulate their motivation in this fashion, they were setting the stage for what would become the third principle for the nature-based approach to AI: intelligence is too complex to design from scratch, so reverse engineer and follow nature's blueprint. In this case, the blueprint called for a neural network.

Rosenblatt's perceptron had a tremendous impact on the nascent field of AI. Throughout the 1960s a growing number of researchers turned their backs on conventional AI algorithms to build perceptrons and similar neural networks in an attempt to expand on these systems' pattern recognition capabilities. Some of these early efforts even resulted in practical applications: the first sophisticated missile navigation system was a neural network modeled after a pigeon's nervous system, and a neural network employed to detect and eliminate phone line echoes is still in use today.

But the greater the enthusiasm and momentum of the neural network movement became, the more dismayed were the "symbolicists" or "cognitionists" of conventional AI. These were people like Minsky, McCarthy, and Feigenbaum who believed that the only sensible way to achieve machine intelligence was to find a formal scheme to represent knowledge and reasoning and then to program this scheme into a digital computer. To these AI leaders, the idea that a machine could create its own representations of knowledge, its own logic, via weighted connections was a dead end.

This schism in the AI community was not merely a battle over which approach was more promising, since both approaches were producing interesting results at the time. This was, rather, a philosophical crusade taken up by those AI researchers who insisted on remaining true to a strict Cartesian view that any good science must be deconstructive in nature; it must break down complex phenomena into clearly delineated theories that allow complete analysis of everything that is going on. In other words, creating intelligence required a solid theory of the mind, a set of equations or formal statements that described the rules and components of intelligence. Neural networkers, on the other hand, seemed to believe that a protomind, if only an extremely primitive one, could be constructed in spite of the fact that the detailed inner workings of the device were not readily analyzable. For this reason, the neural network movement was not merely an annoying distraction to the symbolicists, but rather was perceived by them as a threat to the whole cloth of science. And early AI leaders like Minsky were very concerned that AI be perceived as a solid science.

There were also extremely practical reasons for resenting the neural networkers. The neural network movement was increasingly siphoning off funding and promising young researchers—always a precious commodity—from the classical AI field. Minsky was especially peeved by the growth of neural network research. Early in his career, he had been enthusiastic about neural networks, writing his Princeton thesis on the subject. He met Seymour Papert, with whom he would later found the MIT AI Lab, at a conference when both presented papers on perceptrons. But Minsky, like Papert, soon became convinced that symbolic processing on digital computers was the only way to go in AI, and saw the growing attention being paid to neural networks as a direct threat to his own ambitions. And as if his scientific and practical convictions weren't enough, there was also a personal issue: Minsky and Rosenblatt had been bitter rivals since childhood, when they were classmate prodigies at the Bronx High School of Science. "Minsky couldn't stand the publicity Frank was getting," says one scientist who knew them both.

By the mid-1960s, Minsky and Papert had decided they had seen enough time, money, and enthusiasm sunk into a rival field that they believed wouldn't lead anywhere. The result was a book called *Perceptrons: An Introduction to Computational Geometry*. The book virtually wiped out the field of neural networks overnight.

In *Perceptrons*, Minsky and Papert wisely chose to avoid the tack of objecting to perceptrons on philosophical or even general scientific grounds. Instead, the book reads as an objective exploration of what perceptrons could and couldn't do. The elucidation of perceptrons' capabilities was straightforward, but more startling was the authors' success in proving that the systems, in their simplest forms, were fundamentally incapable of such simple-sounding tasks as differentiating a T-shape from a C-shape or of determining whether an image consisted of an odd or even number of points.

In fact, perceptrons *could* do all of the things that Minsky and Papert said they couldn't, but only if the systems included what are now called "hidden units." Hidden units are a bank of nodes that lie between the bank of input nodes and the bank of output nodes; instead of being connected directly to output nodes, the input nodes are connected to the hidden units and the hidden units are connected to the output nodes.

The importance of having extra layers of nodes between input and output nodes can be seen in the way our brains operate. If our input neurons were directly hard-wired to our output neurons, as with simple perceptrons, there would be a simple, direct, more or less invariable correspondence between the information we took in from the world and the effect this information elicited from us. When we experienced the "input" of a certain piece of music, for example, we would be forced to respond reflexively with the "output" of a predetermined behavior. But a piece of music can evoke many different types of responses from us, such as singing along, blotting it out, or feeling blue; this complexity arises from various types of intermediary processing we perform on the information from our ears: identifying the music, assessing the mood of the music, retrieving memories associated with the music, and so on.

A perceptron or other artificial neural network can perform intermediary processing, too, via its hidden units. These hidden units accept the signals from the input nodes and perform some sort of categorization and reformulation of the information before passing on a signal to the output units. The input nodes are what the neural network "sees," and the output nodes are what the network "does," or "concludes"; in a sense, the hidden nodes represent the way the network thinks about what it sees, the way it finds meaning in the input information, before it acts. If the input is a letter of the alphabet, for example, the hidden units pull out of the images features that are useful for recognition, such as how many curves are in the image, and whether there are more vertical lines than horizontal lines. The hidden units are essentially formulating rules of their own devising that help them with their chores. By adding a layer of complexity to the neural network, hidden units tremendously amplify the breadth and depth of a network's capabilities.

Rosenblatt and others had long been experimenting with hidden units in perceptrons, but no one had yet found a practical method for training such multilayer neural networks. The problem lay with the fact that the extra layer of nodes added a new set of weights to be adjusted. Rosenblatt could use his normal training method to set the weights between the output nodes and the hidden units—that is, increasing the weights to those output nodes that should have fired but didn't, and decreasing the weights to those output nodes that shouldn't have fired but did. But this technique didn't work with the weights between the input nodes and hidden units. The whole point of the hidden layer is to provide the network with the opportunity to come up with its own rules of firing; since the rules aren't predetermined, an observer can't readily determine which hidden units are supposed to fire in response to a particular input.

When Minsky and Papert described the perceptron's inability to perform certain simple tasks, they were focusing on perceptrons without hidden units. As for neural networks *with* hidden units, the book essentially dismissed them—without careful analysis—as being of only marginally more promise and too difficult to train. "We consider it to be an important research problem to elucidate (or reject) our intuitive judgment that the extension is sterile," sniffed Minsky and Papert in the book. This brief statement was the most consequential one of the book. It said, in essence, that the only research project in neural networks worth carrying out was to prove the intuitively obvious (to Minsky and Papert's thinking) point that multilayer neural networks were as big a waste of time as single-layer perceptrons.

Such was Minsky's clout at the time that within a few years of the appearance of *Perceptrons* in the late 1960s, the field of neural networks had become an intellectual wasteland. Virtually all the researchers who had leapt into perceptron research leapt back into symbolic AI. Rosenblatt continued to work on the problem of training multilayered neural networks, making significant progress, but from that point on he had a great deal of difficulty attracting funding and intellectual support; he died in 1970 without ever seeing his work vindicated.

Minsky and Papert claimed, and continue to claim, that the chilling effect their book had on neural network research was unintended and caught them by surprise. As Papert wrote years later: "Did Minsky and I try to kill connectionism? . . . Yes, there was *some* hostility in the energy behind the research reports in *Perceptrons* . . . [but] we did not think of our work as killing [neural networks]; we saw it as a way to understand [them]." Some neural network researchers find such assertions disingenuous, and suggest it is only poetic justice that today it is Minsky's and Papert's work that is largely ignored.

Nevertheless, the damage was done, and by the early 1970s there was little work being carried out on neural networks. As an essay by Hubert and Stuart Dreyfus later put it, "Only an unappreciated few took up the 'important research problem' " that Minsky and Papert had offhandedly thrown down: determining if multilayered neural networks could be trained to escape the limitations of single-layered networks.

Interest in neural networks picked up somewhat in the late 1970s, but the breakthrough that ultimately jump-started a neural network renaissance is often credited to John Hopfield, an MIT physicist-turned-neuroscientist. In 1982, Hopfield started to think about certain similarities between neural networks and collections of atoms. In most materials the atoms, which act something like magnetic compass needles, tend to either twist freely, as in a gas, or else are locked into position via chemical bonds. But in certain exotic materials known as "spin glass" the atoms all twist wildly in response to each other's magnetic fields, finally settling into a stable state—that is, a certain configuration of magnetic orientations—where all the twisting stops. This stable state is the lowest-energy state immediately available to the atoms; to twist into a different state would require added energy, so the atoms tend to fall into this particular set of orientations and stay there.

Hopfield's insight was that the nodes in a multilayer neural network can be described by much the same mathematics as spin glass atoms: given a particular input pattern and an associated output pattern, the weights between the three layers, if allowed to self-adjust, will push each other back and forth like twisting atoms until they finally settle into a stable state analogous to a low-energy state. When this state is achieved, the hidden units are properly connected. By 1984, Hopfield's work had inspired the first functioning, trainable, multilayered neural network. This neural network, invented by Terry Sejnowski and Geoffrey Hinton and called a Bolzmann machine, used a repeated, slight, random reshuffling of weights to gradually approach the lowest energy state.

Bolzmann machines effectively learned to recognize patterns, but the constant reshuffling of connections ate up computer processing power, rendering the model of little practical use. (Virtually all neural network research by this time took place on conventional digital computers with software that simulated the activity of hundreds or even thousands of interacting nodes.) The next leap in the field took care of that problem. Eventually called "back propagation," it was originally hit upon in the 1970s by Paul Werbos, then independently reinvented in the early 1980s by David Parker, and finally fully developed by Hinton, David Rumelhart, and Ronald Williams.

Back propagation is essentially a version of Rosenblatt's perceptron training scheme with an important twist. (In fact, Rosenblatt came heartbreakingly close to stumbling on this method himself, but never quite put it all together.) Training a "back-prop" network begins with providing an initial input pattern, to which the network responds with a random, incorrect output. Adjusting the outermost layer of weights— the weights of the connections between the output layer and the hidden units—is simply a matter of following the perceptron training rule. The hard part is getting at the first layer of weights. To accomplish that adjustment, the network is run *backward*. That is, instead of having the input nodes send a signal to the hidden units followed by the hidden units passing on a signal to the output nodes, the reverse happens: the output nodes send a signal to the hidden units, which relay a signal to the input nodes.

What's more, these backward propagating signals carry the information about how wrong the network's initial "guess" was. In other words, those output nodes that incorrectly fired when they shouldn't have will send back a "weight-weakening" signal that will travel back to the hidden units and then to the input nodes, weakening those weights along the way. Likewise, the output nodes that should have fired but didn't will send back a weight-strengthening signal to adjust the hidden layer of weights. As with perceptrons, very slight adjustments are made in each round to avoid wiping out the effect of previous adjustments, thus typically requiring tens of thousands of rounds of adjustments to complete training. And, as with perceptrons, some sort of supervisor (typically a human, but a conventional computer can play the role) is required to let the network know during training which output nodes haven't fired correctly.

Back propagation worked so well that, like a smoldering dry log, the field of neural networks suddenly roared to life. Theoretically, any computational problem—though practically speaking, any problem that can conveniently be set up as a pattern-recognition task—can be handled by a multilayered back-prop network. Back-prop networks have been employed to recognize speech, to read handwriting, to control robots, to diagnose car problems, and to provide stock-trading advice. Thousands of researchers have thrown themselves into finding new ways to apply the approach.

In fact, the excitement in recent years about back-prop applications is, to some observers, uncomfortably reminiscent of the exhilaration of the AI community when rule-based ("if-then"-style) expert systems became commercially promising in the early 1980s. In that debacle, much of the AI community gave up exploring new lines of research in favor of wringing as much benefit as possible out of what turned out to be a technology of somewhat limited scope and usefulness. Is the same thing happening to the neural network community with back-prop networks?

The problem with back-propagation networks is that while they overcome the specific limitations Minsky and Papert listed in *Perceptrons*, they are still open to the same general criticisms of neural networks that motivated that book and that turned off many AI researchers. Like perceptrons, back-prop networks require absurdly lengthy, human-supervised training sessions; they have no capacity for absorbing information and drawing conclusions on their own. Though they can learn, at an arduously slow rate, to recognize a given set of patterns, they offer no promise of ever being able to accomplish the flexibility and depth of human-like thought. In fact, though back propagation is a handy trick for setting multilayered weights, neurobiological evidence has pretty much ruled out the possibility that brains employ back propagation or anything like it—neurons only transmit signals in one direction, for one thing, and back propagation simply takes too long to be useful to the brain. In other words, back-prop networks might be useful and interesting, but they do not seem to represent any more of a giant leap toward achieving artificial intelligence than expert systems proved to be.

Back propagation has certainly not caused many longtime critics of neural networks to reassess their position. Minsky remains thoroughly unrepentant of his hatchet job in *Perceptrons*. Indeed, in a recent edition of the book, Minsky and Papert add a new essay in which they essentially reaffirm their assertion that neural network research is unproductive. "Little of significance has changed since 1969," they write of the field as of 1988. "The spirit of connectionism seems itself to go somewhat against the grain of analytic rigor." Minsky is more bluntly quoted in a recent book put out by the MIT AI Lab as saying, "The trouble with neural networks is that they are too stupid."

One of neural networks' most incisive critics is MIT's Tomaso Poggio. In addition to his work on machine vision, Poggio spends a fair amount of time writing mathematics-crammed papers that attempt to do to multilayer networks what Minsky and Papert did to perceptrons. "Connectionists predict that the right hardware will spontaneously—perhaps magically—organize itself into a system that is intelligent," he writes in one paper, " . . . [but] many of the networks work only because the necessary computational analysis has been done first."

Unlike most of the back-prop-heavy neural network community, which has remained oddly silent in the face of such comments, Stephen Grossberg loves to respond to neural networks' critics. Unfortunately, Grossberg generally won't allow these responses to be printed, since they typically incorporate a raft of colorful comments and vivid anecdotes that call into question the intellectual and personal integrity of those to whom he is responding. That may seem like unprofessional and even unseemly behavior, but if anyone has earned the right to offer an overly spirited and even vindictive defense of neural networks, it is the fifty-three-year-old Grossberg.

Whereas most neural network researchers came pouring into the field in the 1980s when Hopfield's work and back propagation gained attention, Grossberg is one of the "unappreciated few" who helped found the field and never gave up on it. Working for more than three decades both on theories of the brain and on artificial intelligence, Grossberg has gradually constructed a set of neural-network-based models that seems to explain some of the most baffling problems of thought, and that results in AI systems which see, move, and learn in human-like ways. He is one of the only researchers to even attempt to track the mysteries of thinking, perception, and control all the way down from gross behavior to the level of individual neurons.

What most distinguishes Grossberg's work from back-prop and other types of neural networks is that he has eliminated all the simplifying conditions that enable them to function. Most neural networks function only with "stationary data" and "external control"—meaning they can deal only with a single, unchanging set of patterns, and even then only slowly and under the close supervision of a trainer. A truly brain-like neural network, insists Grossberg, should be "autonomous, fast-learning and adaptive"—that is, it should quickly teach itself how to recognize and deal with a world that is full of surprises.

In fact, Grossberg's insistence on self-learning can be seen as the fourth principle for the nature-based approach to AI: intelligence can't be inserted into a system; it must be developed through the system's interaction with the world around it.

The key to building such a system, he notes, is to tightly incorporate some sort of "feedback loop" that enables the system to evaluate quickly its performance, recognize "mismatches," and take immediate corrective action. Back propagation can't include such a dynamic feedback loop because it is designed to be slowly adjusted to recognize an unchanging set of patterns. "The reason back-prop is so popular is that it opens the loop," says Grossberg, "which makes it a classical system and easy to understand. But in order to open the loop you throw away everything that makes us human. It's orders of magnitude harder to close that loop, but the payoff from doing it is orders of magnitude higher."

Grossberg grew up in Manhattan and graduated valedictorian from Stuyvesant High School. As a freshman at Dartmouth in 1957 he first thought of majoring in philosophy, but his plans changed when he took an introductory course in psychology. "Something hard to explain happened," he recalls. "The data electrified me." Specifically, he was fascinated by his professor's description of a well-known phenomenon called the "bowed serial position curve," which refers to the fact that when people are presented with a list of items to memorize one at a time, they almost always tend to remember best the items at the beginning and end of the list. "If you have a love affair, you remember the beginning and end, and everything else is a blur," he says. "But that's paradoxical. As you're given more and more things to remember it should just get harder and harder. Why would it get easier to remember at the end?" In particular, Grossberg wondered how the brain is able to earmark the end of a list for special memory consideration, when at the time of memorization the brain doesn't even know that the list is about to end. "It's as if when the list ends some sort of effect goes backwards in time in your brain to cause you to remember the items at the end better," he says. "That really fascinated me." It was his first glimpse of the critical role that feedback plays in the human mind.

The "bowed" effect ignited in Grossberg a passion to not merely describe but to account for the workings of the brain—to deconstruct observed psychological phenomena and show how a computational appliance like our brain could produce such behavior. It was not a well-established course of study for an undergraduate, but Grossberg had the very good fortune of coming to the attention of John Kemeny, the Dartmouth math department chairman who had been Einstein's last assistant and is best remembered for having invented the BASIC programming language. Kemeny was just starting a push to get humanities majors to take on a joint major in math, and Grossberg was one of the first students to take the plunge. Soaking up both mathematics and psychology, Grossberg took only until his sophomore year to develop his first neural network model. Grossberg, with typical humility, calls it "the first neural network of the modern era."

By the time he was a senior, Grossberg saw himself as a mathematically oriented psychological theorist, a difficult niche to try to stake out in a field that conferred credibility only on experimentalists. Actually, Grossberg did perform experiments. But they were *"gedanken,"* or thought, experiments—imaginary exercises after the style of Einstein and other physicists who showed that ideas can often be tested without having to get up out of one's chair. "My power was to be able to see into the heart of data," says Grossberg. "And in psychology I was working with the largest, hardest database that humans had ever tried to study. I knew I had to spend every minute of my life trying to explain it. I didn't want to spend time sticking probes into cells."

Kemeny encouraged Grossberg's determination to apply mathematical rigor to psychology, but drew the line when Grossberg wanted to include speculative models of thought in his undergraduate thesis. Kemeny declared he could write about math, or experimental research, but not intuition. Grossberg reluctantly agreed, and again graduated valedictorian. But the minor flap over his thesis was just a mere hint of the trouble heading his way over his bent for theoretical psychomathematics.

Grossberg headed off for graduate school at Stanford, where he attempted to continue his neural network research under the math and psychology departments. But neither department expressed much interest in the endeavor. "The math department said, 'Why would a smart kid be interested in the mind when he could be studying fluid dynamics?'" he recalls. "The psychology department asked, 'Why would he be interested in neural models of behavior when he could be doing statistical sampling?' They just couldn't understand what I was doing." He ended up transferring to Rockefeller University to continue his unusual work; he received his PhD, but still felt ostracized.

He won a faculty position at MIT, but even that school, filled as it was with pioneers in neuroscience and artificial intelligence, didn't know what to make of him and his obsession with neural network models of behavior. His one extended encounter with Papert was so unpleasant as to leave him with his first and only case of hives. Grossberg was ultimately denied tenure there.

Grossberg ended up on the other side of the Charles River at Boston University. BU, too, thought that Grossberg didn't fit in well; but unlike Dartmouth, Stanford, Rockefeller, and MIT, BU decided that the school should accommodate Grossberg rather than the other way around. "I needed to solve an institutional problem," says Grossberg.

"How do you do interdisciplinary work in psychology, neuroscience, physics, math, and computer science? It gradually became clear to me that I had to form an institutional entity that could be a magnet for funding and for people with different types of degrees."

The result was the Center for Adaptive Systems, established by BU in 1981 with Grossberg as director; in 1988 Grossberg additionally became head of a new department of cognitive and neural systems. Graduate students at the center are not merely encouraged to carry on the sort of multidisciplinary study that irritated Grossberg's professors and early colleagues, but are actually formally required to have strong groundings in at least three of the four subjects of math, psychology, computer science, and physics.

Throughout his odyssey, Grossberg never let up on his obsession with developing brain-like, feedback-rich neural networks. Minsky and Papert's book didn't even slow him down; since Grossberg's models were multilayered and autonomous, the perceptron's shortcomings didn't apply to his work. In fact, given the sort of progress Grossberg had been making, a question arises: Why wasn't Grossberg's work more acclaimed during the field's dark ages in the 1970s, when there were few exciting alternatives? For that matter, why do Grossberg's approaches to neural networks today receive far less attention than back propagation?

Grossberg claims that most researchers refuse to follow his line of work because the inclusion of feedback makes the work too complicated for them to understand. "There are two kinds of revolutions," he says. "There's the kind where you have all the math you need, and all that's missing is the insight, the intuitive ideas. That's what Einstein supplied for relativity, and Heisenberg for quantum mechanics. Then there's the kind where both the intuition and the math are missing. That's what we're dealing with here. I and my colleagues here at the Center have had to invent both the insight and the mathematics as we've gone along." He notes that the prospect of having to develop the sort of "nonlinear" mathematics required by feedback models even intimidated such promising nineteenth-century cognitive psychologists as Mach and Helmholtz; to make things easier on themselves they turned to the less complicated realm of physics, where both ended up playing major roles.

Furthermore, because Grossberg's work straddles so many disciplines, researchers often simply see his work as ranging too far outside their realms. Grossberg's intellectual style exacerbates this problem; he seems to enjoy sounding mysterious, even opaque. When lecturing about vision, for example, he makes declarations such as, "All boundaries are invisible," and "Shapes are recognized but not seen"—comments that he can support, but that are at least on first encounter baffling even to many of those who have spent their lives studying vision. Even worse, he is exceptionally fond of co-opting the language of physics at the slightest opportunity: his papers, speeches, and conversation are peppered with such phrases as "global symmetry breaking," "invariance principle," "uncertainty principles," and "complementarity," terms that are near and dear to theoretical physicists but alien and doubtless intimidating to most neural network researchers.

But if Grossberg and his networks haven't received their share of attention, a big reason may be that Grossberg's occasionally outlandish personality has made it easy for others to dismiss him. A first impression of Grossberg does not provide a hint of his capability to pique. He appears gangly and almost elegantly disheveled in his typical outfit of sweater and corduroys, and his gentle eyes amplify a buoyant and almost solicitous charm. But it is not long before one catches a glimpse of a certain churlish arrogance—he refers to it as "bitchiness"—which he seems either unable or disinclined to suppress at the mention of the chilly reception his work has received in much of the AI community.

He complains that others have received credit for his work. The network now known as the Hopfield model, he offers as an example, had been worked out by him decades earlier. "I didn't call any of my work Grossberg nets, I'm nauseated by that sort of thing," he says. "But these cliques are putting their people's names on things I invented." He insists scientists like Rumelhart and James McLelland, largely credited with reviving the field of neural networks in the 1980s, have received far too much acclaim. "People like Rumelhart rode the crest of the wave of interest in neural networks," he says. "He didn't invent anything, but he's a brilliant marketer."

He rails against scientists who indulge in boastfulness and then, moments later, indulges in it himself. "I try not to generate hype, I have no ego about this," he says. "But a lot of these people know from nothing. We are the leaders in the world in revolutionary neural networks, and I am the major pioneer in the world in this field. That tape recorder is going, isn't it?" The reputations of leading researchers drop like flies around him. One famous researcher he names "goes green with envy and hatred every time my name is mentioned." He suggests another is a habitual plagiarizer. Yet another is a "raging egomaniac."

Such comments have earned Grossberg a reputation for vindictiveness and, even worse, self-aggrandizement—the deadliest sin in the science community, eclipsing sheer incompetence or outright dishonesty. "Grossberg spends most of his time citing himself," sneers one leading AI researcher. (In fact, it is typical for most of the citations in a Grossberg paper to be to other Grossberg papers, though that's true of other AI scientists.) His eccentric behavior simply gives many scientists the excuse they seem to be looking for to dismiss his work without having to actually fathom it.

Grossberg admits he has offended people, but even the defense he offers is one that most scientists would find overtly offensive. "I grew up believing that if I were a contemporary of Einstein I would have been at his feet, my face on the floor," he says. "But something is wrong with values today. Being strong isn't fashionable, and that's part of my problem in life. No matter how nice I am about it, when I tell people about the fatal problems with their work I make enemies."

By now, Grossberg is used to the alienation; it may even be what inspires him to stay a pioneer when new scientific ground tends to be broken by young scientists little more than half his age. Perhaps his abrasiveness has even saved him from having to play the role of an AI founder.

ART is in many ways the culmination of Grossberg's intellectual struggle. Developed with Center for Adaptive Systems colleague Gail Carpenter, ART—which stands for "adaptive resonance theory"—arose from Grossberg's longstanding interest in what he calls the "stability-plasticity dilemma." "Stability" in a neural network basically refers to the network's ability to learn new patterns without losing its ability to recall old ones. "You don't want to come in here, meet me, and forget your mother's face," explains Grossberg. "You want to keep learning with confidence that things won't be unselectively blown out of your repertoire."

At the same time, an intelligent neural network should have "plasticity": the ability to accomplish fast learning of new and unexpected patterns. Plasticity is critical to thriving in an environment that's constantly changing and forever throwing out unpredictable events. If it had taken thousands of trials to "train" each primitive human's brain to recognize every new type of danger that arose, from an attack by a saber-toothed tiger to an encounter with an unfamiliar inedible fruit, we surely wouldn't be here today to consider these ideas. Fortunately, our brain effortlessly tosses off phenomenal feats of plasticity. "You can go to a movie one time, see images lasting for tenths of a second and then go home and tell your friends about it for years," notes Grossberg.

But one of the key difficulties in building an artificial neural network with brain-like capabilities is that plasticity tends to be at odds with stability. Learning new patterns quickly requires fast, radical readjusting of weights—but, at least in conventional networks, such a radical retuning can instantly wipe out the ability to recognize previously learned patterns. Back-prop networks epitomize this design trade-off: to avoid catastrophic forgetting, weight adjustments in a back-prop network are kept slight, which prevents these networks from learning new patterns quickly. (Even classical AI researchers have worried about this dilemma, often referring to one form of it as the "frame problem." In essence, the problem posed is how can a system store useful knowledge about the world if the world is constantly changing?)

Our brains, of course, manage to avoid having to choose between plasticity and stability. Grossberg spent years wondering how. "How does the brain switch between its stable and plastic modes without an external teacher?" he says. "How can it learn important events quickly while protecting present knowledge? If the brain were too stable it would be rigid, but if it were too plastic our memories would become chaotic."

Part of the answer, he felt, must lie with the brain's ability to evaluate somehow the relevance or suitability of a new pattern for learning *before* the pattern is learned. If the brain could limit its new learning only to those patterns that provide it with important new knowledge worth saving, then its current store of knowledge wouldn't be overwhelmed by an undifferentiated inflow of useless information. "You want your memory to react quickly," he explains, "but not to irrelevant garbage that will blow out your previous learning."

Such a prelearning evaluation process would explain another characteristic of the brain lacking in back-prop and other artificial neural networks: the ability to categorize an input pattern according to different degrees of specificity. When we see a face, we can make the general decision that it's a human face, the more specific decision that it's a male face, or the even very specific decision that it's Jim Smith's face. This sort of multiple categorization is called "one-to-many mapping," and back-prop and most artificial neural networks can't do it; they can only be trained to provide one output for a given input. The brain, however, seems to be able to consider an input and then categorize it according to different levels of specificity. Most impressive of all, the brain can perform all these marvels without the sort of external supervision required by a back-prop network. Infants learn a great deal about the world around them—such as the difference between a mother and a father, or between a bird and a dog—before they've developed the language skills necessary to be taught by adults.

In his efforts to come up with a model for the brain's solution to the stability-plasticity dilemma, Grossberg considered an idea proposed by Helmholtz over a hundred years earlier. Called "unconscious inference," the theory suggested that the brain tends to perceive patterns in much the way it *expects* to perceive them, based on past learning. In other words, when we learn to recognize certain types of patterns, then new patterns that are somewhat similar will tend to be perceived as familiar, despite slight differences that may exist. If someone speaks the word "stand" to you but a sudden noise drowns out the "a" sound, you'll still usually "hear" the complete word; the brain was *expecting* the "a" sound as part of the aural pattern, so it somehow inserts it into perception.

Grossberg wondered if this phenomenon might not be the key to incorporating some sort of prelearning evaluation scheme into a neural network. Neural networks had always been constructed in a strictly "bottom-up" framework: an input entered the network and was categorized as it moved "up" through layers of nodes, ultimately resulting in an output. But unconscious inference suggested there was also a "top-down" element to the brain, so that not only did input determine output but output could influence input; in other words, the higher-level layers of processing of the network, where recognition occurred, was somehow affecting the perceived input so it seemed more familiar than it otherwise might.

Performing this trick, decided Grossberg, was related to the process of paying attention. By paying attention to the important similarities between a new pattern and a learned familiar one, and by ignoring minor differences, the brain could recognize a pattern without being forced to learn the differences—that is, to change its weights because of minor and presumably irrelevant elements of mismatch. If you were looking for a friend in a crowd you probably wouldn't be thrown off by his wearing a Band-Aid on his chin; for the brief instant it took to make the recognition, your brain would pay attention to his features and ignore the Band-Aid (though once the recognition is made, you might start focusing on the Band-Aid out of curiosity).

Of course, if there were too many significant differences between a pattern and an expectation of it, then the brain could decide it doesn't have a match and categorize the new pattern differently. For example, if you spotted someone who looked vaguely like your friend, but on closer inspection realized this person had a wider nose and a smaller mouth, your brain would deem it a mismatch and forget it, or might even create a new category of "people who look like my friend" and remember the wider nose and smaller mouth. In other words, learning would take place only when a pattern was properly categorized and would focus on the most salient aspects of the pattern. Pattern variations that were slight or unimportant—for example, a piece of popcorn on your friend's mouth—could be filtered out before learning occurred. Thus the brain can recognize familiar patterns and learn new ones without cluttering itself up with the cloud of noisy, ordinary variations that characterize the world around us. Plasticity without instability.

• • •

It was around this line of thinking that Grossberg and Carpenter conceived the ART neural network in 1987. An ART neural network consists of four main components: a short-term memory to evaluate input patterns, a long-term memory to store categories, an attentional subsystem to focus attention on important features of input patterns, and a reorienting subsystem to keep the long-term memory from learning irrelevant patterns. When ART begins learning, its long-term memory is empty. The first input pattern—say, for example, an image of a four-door car—is presented, and the raw data go to the short-term memory, which passes the data on up through a set of randomly adjusted weights to the nodes that make up the long-term memory. (This is the "bottom-up" information flow.) Here, the nodes "compete" to match the input pattern; that is, the set of nodes in long-term memory that fire in a pattern most closely resembling the weighted input pattern becomes the "winner." This winning long-term memory pattern is essentially ART's first "guess" at a matching representation of the input.

But unlike other neural networks, ART doesn't depend on an outside supervisor to tell it if its guess was right. Instead, ART assesses the guess itself by relaying it back down to the short-term memory for comparison to the input. (This is the "top-down" information flow.) At this point, the attentional subsystem causes those parts of the original input pattern and the guess pattern that match to stand out, and suppresses those parts of the two patterns that don't match. In essence, ART is "paying attention" to the "familiar" parts of the input pattern—that is, the parts that match the pattern in its long-term memory—and "ignoring" the unfamiliar parts. This step essentially creates a new input pattern that consists of a blend of what ART "expected" to see and what it is actually seeing.

Now ART's reorienting subsystem takes a look at this combination bottom-up and top-down pattern to determine if it is substantial enough to declare a match. The reorienting subsystem's standards for making this decision are adjustable; ART can be set to be extremely finicky about making a match, requiring that the patterns be nearly identical, or it can be generous, requiring only that there be minimal areas of overlap. A generous setting would be useful for, say, telling whether or not a pattern represents a face (as opposed to an object that isn't a face); a finicky setting would be useful for determining whose face has been input.

If the overlap isn't good enough, the reorienting subsystem decides ART doesn't have a match, and it starts the process all over again by causing the original input pattern to be sent back up to long-term memory. But first, the reorienting subsystem removes from the competition those nodes that composed the first guess. In addition, some of the long-term memory weights are reshuffled in the hopes that a better guess will result; in a sense, ART is considering new possibilities. Again, a best guess pattern is selected and sent down to short-term memory for evaluation via the attention-focusing routine. If that guess doesn't measure up to ART's standards, the process is repeated yet again, and so on until a good-enough match is produced.

When long-term memory finally produces a pattern close enough to the original input pattern, the reorienting subsystem remains quiet. That leaves the blended input/expectation pattern active in short-term memory, and the persisting presence of this signal starts to change the weights in the winning long-term memory nodes, which results in an even closer match in short-term memory, which further changes the weights in long-term memory, which results in an even closer match, and so forth. At this point, ART has entered a state of "resonance"—a sort of runaway feedback loop that completely locks ART's attention on the match between the two patterns. This is the "whoomph" of recognition you feel when you finally see your friend's face in the crowd; your eyes lock on the face and your brain is temporarily awash in the sense that you've made a match.

It is only during this state of resonance, and not until then, that ART is free to "learn"—that is, to adjust the weights of the winning nodes in its long-term memory to more closely match the input pattern. The winning nodes, for example, may have produced a pattern that was at first only vaguely car-like in shape; but when resonance and its attendant learning occur, the weights are adjusted so that next time an image of a four-door car is input the nodes will immediately produce a close match. The process is akin to what happens if you'd recognized your friend and had noticed he had grown a mustache since you'd last seen him. The mustache might have delayed recognition for a moment this time around, but your brain will make the necessary adjustments—it will "learn" the mustache—so that the next time the mustache won't slow recognition (and may even speed it up).

Imagine that the next pattern presented to ART is an image of a truck. Again this input pattern is brought into short-term memory, and again a search takes place in long-term memory for the best match. Suppose the "car" nodes win again. The car pattern is then brought down to short-term memory as ART's "expectation" of what an input pattern that fits this category should look like. ART then focuses its attention on the similarities between the truck and car patterns, and ignores the differences. If ART is set to a low degree of specificity, it may go into resonance with the "car" nodes, calling it a match; when it learns the new pattern, the car nodes will then become more truck-like, so that the nodes may ultimately come to represent a general "wheeled vehicles" category. If ART is set to a high degree of specificity, then it may reject the match with the car nodes and quickly come up with nodes that are more truck-like—without making any changes in its memory of cars. From then on, cars and trucks will go into separate categories.

Thus, like the brain, ART offers both plasticity and stability: it can immediately learn a new pattern without forgetting old ones. It also provides one-to-many mapping, allowing a given pattern to be placed in general or specific categories. Most significantly, ART organizes its own categories, performs its own guess evaluation, and carries out its own learning. "The art of ART," says Grossberg, "is that it will find the best guess even though no one tells it how to do it. It's a self-organizing system."

Grossberg contends ART's top-down feedback system serves much the same function as the section of the brain known as the hippocampus. Without its feedback system, ART can't build new long-term memories; it can recognize patterns that are already established, but it can't learn new ones. In the same way, patients such as severe alcoholics known to suffer hippocampal damage can often tell what is going on around them minute to minute, but can't recall any of these events an hour later.

Even more intriguing, recordings of electrical activity in the brain suggest there are specific "waves" of activity during recognition and memory formation that seem to correspond to the sequence of signals employed by ART. Neurobiologists have noted that a brainwave known as "N200" accompanies a perception in the part of the brain that provides short-term memory, answered by a "PN" wave from long-term memory, apparently corresponding to ART's bottom-up and top-down signals; when the brain is struggling with recognition it issues "P120" waves to short-term memory followed by "P300" waves to long-term memory, apparently corresponding to ART's mismatch and reorienting signals. In fact, this might in a sense be considered a case of life imitating ART, since ART was established before the pattern of interaction of these brainwaves was identified. Grossberg claims this correspondence suggests ART does indeed represent a valid model of at least one aspect of the brain's operation. "ART was making a prediction about the learning of top-down expectations in the brain," he says, "and these correlations were observed. This is a more subtle domain of neuropsychological investigation, a domain that other cognitive theories cannot yet penetrate."

ART is also proving its worth as an effective and fast-learning pattern recognizer in practical applications such as letter or image recognition and database searching. Grossberg claims ART generally blows away the competition: in applications for which back-prop networks require 20,000 supervised training rounds, he says, ART teaches itself in 5 rounds, and performs with higher accuracy—in some cases with accuracy rivaling human performance, a feat back-prop and other neural networks almost never achieve. (Not surprisingly, some researchers dispute such claims, suggesting Grossberg's systems work better in theory than they do in practice.)

ART is now beginning to find its way into commercial applications. Boeing, for example, is incorporating ART into an extraordinarily ambitious parts identification plan now under development. When completed, the system will catalog the design specs of some 16 million aircraft parts, allowing the company's engineers to enter in the design specs of a proposed new part and get back information on the closest existing part; that way Boeing can modify the existing part instead of having to manufacture the new part from scratch. (In this and other applications calling for a single, objective, "right" answer, ART can be trained in a "supervised" mode in which a human can essentially veto ART's decisions, ensuring that ART does not end up categorizing patterns according to criteria that are irrelevant to the application.)

In another application, a Nevada medical center is developing an ART-based system to categorize the conditions that relate to the length of patient stay. The system can then predict a patient's length of stay based on the patient's history, current status, and course of treatment. Of particular importance to hospitals is the fact that ART, unlike back-prop and most other neural networks, can produce a list of "rules" indicating how it makes its decisions; that is, ART can show which features of a pattern it has learned to pay attention to when making a particular categorization. The Nevada medical center, for example, could then use these rules to determine which types of treatments result in which types of patients being released more quickly.

Could ART, or something like it, be developed to the point where it can replicate more complex, higher-level arenas of human thought? Could ART learn to converse, give advice, provide companionship? Grossberg claims ART could indeed provide the basis for such a system; the idea, he says, would be to build an ART out of ARTs. He notes that the brain functions on many levels: rather than simply accepting input from the external world and directly converting it to high-level thoughts, the brain sends the information through some thirty increasingly high-level processing steps. Thus when the brain receives the streaks of light and color the eye pulls from an image of an ice cream cone, it first extracts from the pattern salient features such as the triangular outline of the cone and the surface texture of the ice cream, then it assembles the image, then it identifies the image as an ice cream cone, then it pulls up memories and knowledge about ice cream cones—they taste good, they're cooling, they're fattening, they're messy—and only then puts together a high-level thought: I probably shouldn't, but I'm going to get an ice cream cone.

In the same way, says Grossberg, a number of ARTS could theoretically be stacked together to create a sort of superART capable of higher-level thinking. "You'd have a hierarchy of systems," he explains, "where each level categorizes and abstracts the information in the level below, and is being categorized and abstracted by the level above. The lower levels would learn quickly, while higher levels would be relatively invariant, changing only very slowly." In fact, notes Grossberg, this idea of putting together levels of increasing abstraction is not new to AI; the difference, he says, is that a superART would be self-organizing.

Grossberg is heading his work in that direction. He and associates are working on linking together various ART-based and other brain-like models he has developed over the years. One model, for example, simulates the brain's long-baffling ability to perceive images by recognizing object borders and then filling in colors, even on moving objects. Combined with ART's ability to focus attention and overlay expectation on input patterns, Grossberg's model seems to provide a surprisingly good imitation of natural vision and image recognition—and one that a growing number of neuropsychologists are incorporating into their theories.

Grossberg also works on the problem of motor control—that is, the ability of humans and animals to activate their muscles in such a way as to perform phenomenally precise, high-speed tasks. In particular, Grossberg was fascinated by the fact that even a child can unerringly enact a complex reaching motion on a first try. That capability, which we perform so effortlessly and take so completely for granted, requires a raft of faculties: we must construct and store a map of the world around us, this map must be translated into a knowledge of the distances and directions of objects in relation to our bodies, this knowledge must be translated into a path of motion from our hands to the object we want to grab, the path must be translated into signals that control a complex symphony of multiple muscle motions of varying strengths and speeds, and then the resulting motion must be precisely monitored for errors and corrected on the fly as the hand heads toward the object.

Grossberg has developed a series of models that he claims performs every one of these tricks. "These models link two separate worlds in robotics—the high-level world of planning and maps and the low-level world of reaction, force control, and precision," he explains. "The problem is how do you get the ideal world of planning to work with the world of movement and forces. Or, in other words, how do Plato and Newton talk to each other in the brain?" In Grossberg's system the world of Plato is handled by components analogous to parts of the cortex in the human brain, and the world of Newton by components analogous to the cerebellum and the spinal cord. As with ART, the components resemble their brain counterparts in their nonfunctioning as well as their functioning states: when the cerebellum-like component is disabled, for example, the system behaves like a Parkinson's disease patient, sending out spurious motor signals.

As Grossberg and colleagues work on integrating these various systems, their computer simulations are becoming more and more competent at a broadening array of tasks. In some cases, the projects are going beyond computer simulations to robots that are controlled by neural networks based largely on Grossberg's various brain-like modules; some of these robots tackle tasks that are well beyond the capabilities of traditional AI. Allen Waxman at the MIT Lincoln Labs, for example, has constructed a robot that roams around a room looking for particular objects and avoiding others. And Eric Schwartz at the Center for Adaptive Systems has built a tiny robot with rotating eyeballs that attempts to recognize a license plate on a moving car.

To get to the next step—systems that could teach themselves human-like behavior—Grossberg will have to overcome a slew of mysteries as to how the brain deals with the confusion of a complex, fast-changing, and often hostile world. One is what he calls the "adaptive timing problem": the way the brain decides whether or not to change its behavior when the consequences of that behavior are delayed, or sometimes altogether absent. "In the real world we're not always lucky enough to have a teacher, a rational, compassionate God, who punishes and rewards in a clear way," he says. "Most of the time nothing happens when we act; we have to learn by the fact that our expectations are disconfirmed."

The problem, then, is getting a system to learn that if an anticipated rewarding consequence doesn't follow a certain behavior, then a new behavior must be attempted. On the other hand, the system has to learn to wait a reasonable length of time for the anticipated consequence before deciding it's time to move on to a new behavior. "We'd have to pay a huge price if our sensory and learning systems were too fast, and all hell broke loose with every mismatch," explains Grossberg. "We'd all be roaming the streets with our tongues dragging on the ground. The trick is to know when to shift attentional focus and start exploring new behaviors." People such as war veterans who have experienced traumatic stress often suffer from a breakdown in this mechanism, says Grossberg, resulting in their being stuck in a feedback loop where their attention locks on anxiety-producing events, which triggers inappropriate behavior, which creates more anxiety, which triggers more inappropriate behavior, and so on.

To get around the adaptive timing problem, Grossberg has designed neural network models that build up brief "histories" of their actions and the resulting events. The models, which have already been incorporated into a robot, can then anticipate what is likely to happen within what time-frames, and thus can wait an appropriate amount of time before losing old learning and exploring new behaviors. "Five years ago no one even realized that adaptive timing was a problem in learning," says Grossberg. "Half of what we have to do here is figure out what the problems *are*."

Of course, even if Grossberg is able to create a neural network model that achieves higher-level modes of thought, such efforts by themselves will not result in a system that provides anywhere near the brain's level of performance. Such a system, for example, would have to tackle the issue of processing speed: the brain can recognize an entire roomful of objects at a glance, while the fastest artificial image-recognition system running on a supercomputer chokes for long seconds on a single object. A brain-like system would have to be phenomenally precise: the brain can make out a whisper in a crowded restaurant, while ART-like networks require relatively favorable conditions to discriminate and identify simple patterns. And it would have to span a wide array of capabilities: the brain can process the huge flow of information from five senses, control hundreds of muscles acting simultaneously, negotiate complex and rapidly changing surroundings, remember and learn from decades of experiences, create works of art, and solve problems through logic and sudden insight; the best ART system, or any other system, can't approach the brain in any one of these areas, let alone handle all of them at once.

Increasingly, researchers following the nature-based approach to AI are recognizing that if a neural network is going to even approach such proficiency it will have to be imbued not merely with a better software design but with some sort of hardware advantage as well. The brain, after all, is not characterized simply by a clever algorithm that might as easily run on an IBM PC; it is an extraordinary physical computing device in its own right. Imitating it on a hardware level is an entirely different sort of challenge.

4. BRAINWARE

*Aimed by us is futuristic humane machines
wherein human level electronic intelligence
and nerve system are combined to machines
of ultra-precision capabilities.*
—A BROCHURE FROM MATSUSHITA
RESEARCH INSTITUTE, TOKYO

Sipping green tea and listening to the Carpenters in his cramped Yokohama office, the polite, carefully spoken Masuo Aizawa doesn't look like the sort of scientist whose work would raise eyebrows. Nor does a cursory examination of the fruit of his labors give cause to question this assessment: it is a glass slide floating at the bottom of a plastic dish filled with a clear liquid. As it turns out, the slide is an electronic chip of sorts, though a peek under the microscope suggests it is a crude one by any standards. Instead of the intricate paths and byways of modern chips, this one offers plain, broad stripes; where conventional chips are adorned with millions of infinitesimal transistors, Aizawa's seems to have been splattered with mud.

But appearances are misleading. The spindly blobs on Aizawa's chip aren't defects, but living neural cells that have been custom-grown into the precursor of a biological electronic circuit—the first step, says the forty-eight-year-old Aizawa, toward the neuron-by-neuron construction of a semiartificial brain. Aizawa and a number of other researchers, a disproportionate percentage of whom are Japanese, believe it may prove easier to build intelligent machines out of living cells than it will be to mimic the functions of such cells with semiconductor technology. "Perhaps this is just a faraway dream," he says, chuckling. "But we are approaching it in steps."

A good first step in considering ways to implement a high-performance neural network is to appreciate the properties of the cells of which brains are composed. One of the most impressive things about neurons, at least at first glance, is the fact that there are so many of them—about 100 billion, or as many as there are stars in our galaxy, in the 3.5 pounds of gray matter sealed in the skull of a typical adult human. These neurons are connected by a million billion synapses, and collectively fire 10 million billion times per second, while consuming less power than an ordinary light bulb.

There is no exact or generally accepted way to compare the processing power of a brain and a conventional computer, but AI researcher Hans Morjavik has calculated that a desktop computer roughly packs the calculating punch of a snail, while a Cray 2, one of the fastest supercomputers ever built, barely matches the brainpower of a small rodent. Were it possible to build a computer of a capacity roughly equivalent to that of a human brain with current semiconductor technology, the computer would require 100 megawatts of power, enough to light a town.

Even then, such a computer wouldn't be very suitable for running a neural network. Conventional computers process information "serially," performing one simple task, such as adding two numbers, at a time. Neural networks, in contrast, call for vast amounts of "parallel" processing capabilities that can deal with the simultaneous firings of many nodes. Conventional computers can simulate the operations of a neural network in software, essentially by running a program representing a node over and over again until the software has accounted for the activity of each node in the network over a frozen slice of time; then it repeats the process all over again to advance the network one round of firings into the future. Clearly, this is a clumsy and extraordinarily inefficient way to run a neural network, and it is only due to the ability of modern computers to churn through tens of millions of serial instructions per second that such simulated networks are only millions of times slower, and not trillions of times slower, than a bird brain.

The first perceptrons were true hardware neural networks, in that each node was actually a physically separate, albeit simple, electronic circuit capable of firing in response to signals from other nodes. But these early hardware neural networks consisted of only a few hundred nodes, and the steep costs of custom-designing and building larger hardware neural networks sent researchers fleeing to conventional computers in the 1960s. Development of true neural network computers essentially ceased for two decades, with the exception of the occasional project for the Department of Defense, which has long been interested in applying neural networks to such pattern-recognition applications as recognizing the audio profiles of enemy submarines. Thus the most powerful neural network computer built through the 1980s was the DoD-commissioned TRW Mark III neurocomputer, which consisted of one million synapses—impressive for a manufactured device, but about one ten-thousandth the capacity of a housefly.

The computer industry had always all but ignored heavily parallel processing, because virtually all practical software had been written for serial computers, and relatively few computer users (other than neural network researchers) were willing to go through the costly task of converting their software. Only recently has the computer industry begun to pay serious attention to parallel processing, largely because of the huge gains in processing power per dollar that a few computer companies have achieved with parallel designs.

The clear leader among such companies is Thinking Machines, founded by Danny Hillis when he was an AI graduate student at MIT. Though less than a decade old, Thinking Machines had revenues of about $100 million in 1992; its headquarters, to which Hillis sometimes commutes in an immaculately restored antique fire engine, has helped establish a minirenaissance in Kendall Square, that area of Cambridge near MIT that was built up and then deserted by the brief-lived expert systems industry. The gleaming cubes known as the CM-5, Thinking Machines' most advanced computer, contain as many as 1,024 physically separate processors that can simultaneously work on different parts of a problem and then communicate to combine their efforts into a single answer. CM-5s have begun to turn up in the data centers of retailers like Kmart and other large corporations, but no one lusts after these machines more than neural network researchers. Though the CM-5 lacks the huge number of nodes and the highly flexible connectivity needed to qualify as a true hardware neural network, its parallelism can simulate such network orders of magnitude faster than can a conventional computer. It is no coincidence that the leading manufacturer of parallel processing machines was founded by an AI scientist.

With a price tag of as much as $30 million, a CM-5 is wishful thinking for most neural network researchers. But there are other hopes for these scientists. One is that better neural-network-simulating software, combined with ever faster and cheaper serial computers, will reduce the performance gap between hardware and simulated neural networks. One new node-simulating algorithm, for example, is claimed by its designer to allow a desktop computer to rapidly simulate over one million neurons and 40 million synapses.

But most researchers believe that constructing a neural network with a capacity and speed even approaching that of the human brain will require a true hardware implementation. This conviction makes up a fifth principle of nature-based AI: intelligence can't be readily divorced from its physical substrate. Constructing a machine that thinks is a hardware as well as a software problem.

One promising and affordable avenue toward a hardware solution is the recent appearance of "neural network chips." Like the microprocessor and memory chips in conventional computers, these postage-stamp-size silicon wafers are etched with millions of microscopic wires and transistors. The difference, of course, is that the wires and transistors in a neural network chip are designed to serve as neurons and synapses. Chips have two significant advantages over larger processors: they can be cheaply produced—as little as tens of dollars in large enough quantities—and the signals sent within chips travel as little as millionths of an inch instead of several feet as in supercomputers, making them blindingly fast. It is for these reasons that the computer industry has been relentlessly moving toward machines based on standard "microprocessor" chips. (In fact, even Thinking Machines' processors are all built around the same chips used in desktop computers.) The neural network community is likely to do the same.

Most of the neural network chips that have emerged so far are "general purpose" chips—that is, chips with easily adjusted synapses to suit a wide variety of neural network applications. One such chip, under development by Stanford researchers David Stork, James Burr, and Michael Murray, is a Bolzmann machine neural network that promises to be the fastest pattern-learning machine ever built. "This chip is going to jump things way up in terms of the kinds of problems you can solve with neural network," boasts Stork, who has also written a neural network program that enables a camera-equipped computer to read lips.

Stork's chip will have stiff competition. At least one company, called Corticon, has been formed expressly for the purpose of developing and marketing neural network chips. It has already introduced a first chip that can be linked into networks of several thousand neurons. And AT&T Bell Labs researcher H. P. Graf has developed an experimental chip that comprises 256 neurons connected by up to 32,000 synapses whose weights can be changed at the rate of 320 billion per second. Though Graf's neurons are limited in the number of signals they can receive, up to 8 neurons at a time can be combined on the chip into "superneurons" with each handling up to 1,000 connections. The chips can be plugged into a standard computer to allow combining neural network computing with more conventional computing.

Of course, such chips are still a long way off from providing the neural capacity or connectivity of the human brain. But even a chip that could match the number of neurons and synapses in the human brain still would not in itself open the door to creating a neural network of brain-like capabilities. That's because artificial neurons and synapses are far simpler than the real thing. Artificial neurons fire if they receive strong enough signals from the neurons to which they are connected, and artificial synapses can be strengthened or weakened according to simple rules. But among biological neurons the mechanisms for determining when and how to fire and for changing the strengths of synapses are so varied and complex that researchers have only begun to identify them, never mind the details of how these mechanisms work together to define neural circuits. In effect, individual neurons are not so much like the relatively simple switches with which they are represented in artificial neural networks as they are fully functioning computers in their own right. "Artificial neurons are greatly simplified caricatures of biological neurons," contends David Tank, an AT&T Bell Labs researcher who works with both types of neurons.

The most basic functioning of neurons is fairly well known. As with all cells, neurons' exteriors consist of a membrane that keeps the contents of the neurons intact, and that prevents most substances from leaking into the cell. In their normal, or "rest," state, neurons retain inside them a number of negatively charged molecules that enable the cell interior to maintain a voltage that is a modest 70 millivolts less than the outside of the cell. When a neuron starts to receive "excitatory" signals originating from other neurons along its many dendrites, "channels" in its membrane open up and allow an inflow of positively charged sodium atoms. When the sudden accumulation of positive charge raises the voltage of the cell interior to a maximum of about 100 millivolts, the cell fires: that is, the excess charge is carried away from the cell into its single axon. This "pulse" of charge travels along the axon at a speed of about 300 feet per second, or 200 miles per hour. That may seem fast, but in wire an electrical pulse travels at near light speed, or 186,000 miles per *second*. Thus the brain's astounding processing speed is clearly not due to the inherent quickness of its components. In fact, an ordinary transistor switches about one million times faster than a neuron fires.

Unlike the nodes in most artificial neural networks, real neurons aren't limited to communicating in an on/off, or fire/no-fire, fashion. They also send information in the form of the *rate* of firing. In other words, a neuron firing twenty times per second will have one effect on the neurons receiving its signal, while the same neuron firing fifty times per second could have an entirely different effect. (It takes a typical neuron about one two-hundredth of a second to "recharge" after firing, resulting in a maximum firing rate of about 200 times per second.) The relationship between the rates of the incoming signals and the rate of the resulting outgoing signal can be quite complex, and can differ from neuron to neuron. To make matters even more confusing, the relationship between the incoming and outgoing signal rates for a neuron can be entirely different for "inhibitory" signals—signals that make a neuron less likely to fire. In some cases, for example, an increasing inhibitory signal rate can make a neuron *more* likely to fire.

What's more, the properties of a single neuron can significantly fluctuate in response to changes in incoming signal rates, as the neuron apparently "tunes" itself to other neurons. Indeed, the neuron's ability to seemingly mold itself to varying conditions led neuroscientist Harry Klopf, who is with the Air Force's Wright Aeronautical Laboratories, to develop a theory of intelligence based on the notion of "hedonistic" neurons—that is, that neurons "seek" excitatory signals and "avoid" inhibitory signals. Claims Klopf: "Intelligent brain function can perhaps only be understood if it is assumed that the single neuron is a goal-seeking system in its own right."

The idea of a hedonistic neuron, which envisions intelligence as a sort of mass Pavlovian conditioning phenomenon, is not one to which many researchers fully subscribe. But virtually all brain scientists agree that neurons are extraordinarily adaptive, thus imbuing the brain with a richer set of capabilities than could a network of interacting simple switches. "What we are finding," says California Institute of Technology neuroscientist Christof Koch, "is that, depending on the neural network around it, a single neuron can exhibit very different properties."

In view of all these complexities, it is not surprising that the functioning of nodes in artificial neural networks tends to be extremely crude in comparison to neurons. What's more, the synapse-adjustment mechanisms in such networks are also highly oversimplistic. The brain incorporates myriad complicated mechanisms for increasing and decreasing signal strengths at synapses, including the activities of dozens of so-called neurotransmitters and neuromodulators—chemicals released by various cells that can strengthen or weaken the signal between two neurons, often at the command of yet other neurons.

Obviously, simulating such a dizzying array of mechanisms would be a lot easier if scientists were able to at least identify most of them. But so far they have proven elusive. One of the most basic mechanisms in a neuron, for example, is the opening and closing of the "ion channels" that allow charged molecules and atoms to come in or out of the cell, but even these are obscure. "There are so many of these ion channels and their interrelation is so complicated," says University of Wisconsin neurobiologist William Lytton, "that it is not possible to produce their effects."

The dearth of information about neuronal function certainly does not result from a lack of trying. The neurons of thousands of different creatures—and particularly those of monkeys, cats, rabbits, and rodents—have been studied in detail. There are researchers who specialize in the neurons in lobster stomachs, and those in the lamprey eel's spinal cord. Neurons are examined both on microscope slides and, often under circumstances that would strike most of us as grotesque, in living animals. To study the properties of vision-related neurons, for example, Stanford neurobiologist Carla Shatz has performed a series of startling experiments in which she operates on living fetal kittens in the womb. Working in the dark through a small slit in the anesthetized mother cat's belly, Shatz removes one eye of the fetus while leaving it connected to the optic nerve, and attaches an electronic chip over the field of neurons at the back of the eye; she then cuts open the eye and exposes the inside of it to various light patterns, allowing the chip to register the signals sent by the neurons to the optic nerve.

Such research has provided a wealth of information, but the more that is learned about the brain, the more it seems there remains to be learned. Each discovery raises new questions that could—and often do—consume the careers of dozens of neurobiologists, and further highlights the vast scope of the challenge to those that would construct brain-like artificial neural networks.

The brain doesn't employ generic neurons. Neurons involved in vision are different in form and function from those involved in hearing. Some neurons specialize in receiving signals from the outside world, some in passing on signals to other parts of the brain, some in transforming information. In the hippocampus, so-called Hilar neurons seem to exist primarily to keep other nearby neurons firing in an orderly fashion; when Hilar neurons are damaged, the brain often suffers the disorganized storm of neural firings that constitutes an epileptic seizure. Even within patches of seemingly identical neurons, a probing of the cells' DNA reveals differences in genetic makeup.

Most of these differentiations and specializations are complete mysteries. Researchers suspect that in many cases they may be related to neurons' tendency to be especially sensitive to certain signal patterns. It has been known for many years, for example, that in the human brain there is a large number of neurons that fire only in response to a human face; in addition, many of these neurons seem to specialize in recognizing certain features of faces, such as eyes or mouths. Some researchers believe the brain assembles even the most complex images out of a simple "alphabet" of all-purpose simple shapes, or "icons," recognized by neurons that specialize in these icons. What's more, neurons of related specialties seem to organize themselves into small slab-like structures the size of a paper match body. These "columns" can be observed throughout the brain, though their purpose isn't understood.

Neurons are constantly changing. The growth and movement of axons and dendrites are particularly pronounced during fetal development, of course, but they continue to wander—according to schemes that are not understood—throughout the brain's life. Synapses can be quickly but temporarily altered by proteins that act and decay in moments, perhaps accounting for short-term memory. Or they can be affected, many researchers believe, by genetic changes that can last a lifetime, explaining why, for example, most of us never forget our names.

Sleep is clearly a critical mode of brain operation, yet there is little agreement among researchers as to its exact purpose. (Some researchers, for example, have proposed that the brain uses sleep to reinforce the learning of critical information encountered during the day by strengthening synapses; others claim it's the brain's way of "erasing" information that hinders thought patterns.) Ongoing neuronal death, or "pruning," seems to play an important role as well: human fetuses lose half the neurons in the back of their eye by birth, as the brain apparently streamlines and focuses its circuitry. Exceptionally intelligent people may be those not with extra neurons and connections, but those whose brain does a better job of ridding itself of unnecessary neurons and connections.

Some of these neural behaviors can be mimicked, in a fashion, in artificial neural networks. For example, Stanford's David Stork and graduate student Babak Hassibi have designed a technique they call the "Optimal Brain Surgeon" for pruning the nodes and connections of a neural network, typically achieving a tenfold improvement in performance over unpruned networks. But such efforts are generally not presumed to even approximate the effectiveness, and perhaps not even the purpose, of the biological versions. And researchers usually don't even attempt to imitate neural behaviors, either because the behaviors' purposes aren't understood, as is the case with sleep, or the means for imitating them remains elusive, as is the case with fetal axon and dendrite self-organization.

Of course, one can reasonably assert that the fact that such neural mechanisms exist in the brain does not necessarily mean they are critical to an artificial neural network. Indeed, the creed of traditional AI is that life is a big evolutionary kluge—a hack—and it would be a waste of time to imitate it. Though this extreme (but still widely held) view is vehemently rejected by nature-based AI researchers, even they don't all agree on the appropriate degree of biological imitation. Still, many researchers who have carefully considered mechanisms of both biological and artificial neural networks wonder if it is possible to throw out the bathwater of biological complexity without throwing out the baby. "Why do real neural circuits have such complex physical mechanisms underlying the simplest cellular properties," muses AT&T's Bell Labs' David Tank, "and why are the patterns of connectivity far from random in the simpler circuits that have been studied?"

One of Tank's main goals is to identify the "computational logic" in pieces of biological neural networks—what he refers to as "circuits"—and to determine ways to divorce the circuit's functions from the physical mechanisms. In other words, Tank wants to figure out which aspects of the brain's implementation of intelligence are worth translating onto a silicon chip, and how that translation might be accomplished. Tank has encountered formidable difficulties in achieving such translations even for some of the best-known, simplest neural circuits, and contends that not a single such circuit has ever even been fully understood. Still, he has been able to construct reasonable electronic imitations of the circuits that, for example, allow barn owls to determine the location of sounds, and bats to perform echo location, while another Bell Labs researcher has assembled a device that fairly duplicates the three-neuron swim circuit of the mollusc Tritonia.

Given the fact that such relatively trivial neural circuits have proven so difficult to emulate in silicon, it may seem almost laughable to hear California Institute of Technology researcher Carver Mead proclaim that silicon versions of large portions of the human brain will be operating within a decade or two. But while most scientists and computer engineers are indeed skeptical of this prediction, few are willing to risk laughing at Mead. The last time Mead provoked ridicule was when he predicted in the early 1970s that silicon chips would incorporate one million transistors within a decade. Not only did Mead prove right, but he personally developed much of the theory and design that allowed reaching that goal. Since then, Mead has been working on silicon implementations of biologically motivated neural networks, especially those related to vision. "Computing wasn't much closer to real-time vision than it was in the 1950s," he says. "But the brain can do it. So I realized we must be missing something."

With his angular face, piercing eyes, and goatee, Mead looks more like a medieval wizard than an electronics wizard. Lately, he is often seen in the company of a man who could be, to judge by appearances, Mead's converse: that is Frederico Faggin, the refined, nattily dressed president of Synaptics, a company formed to develop and market Mead's design for a neural network vision chip. Faggin, who designed the first Intel microprocessor chip, plays the pragmatist to Mead's visionary. "I believe in the principle of observing nature as a source of incredible designs," he says in modestly Italian-accented English, "but when it comes to solving problems you must apply known technology within the boundaries of cost, power dissipation, and physical size."

The result of this collaboration is the first commercially available chip to be directly modeled on a biological neural circuit. Dubbed a "silicon retina," the chip gathers an array of light-sensitive nodes in a circular arrangement resembling that of the neurons behind the eye. Like visual neurons, these nodes detect differences in brightness between different points on an image, and assemble these differences into edges and other features allowing quick recognition. As a result, the chip is capable of reading the bank numbers of up to ten checks per second, a far faster rate than is provided by conventional technology. "This isn't the kind of approach that an engineer would normally think of taking to this problem," says Mead, "because most of the things the brain does aren't analogous to the things we have engineering disciplines for. That's why building this sort of chip is so hard, and that's why it's so exciting."

Mead's work has inspired a number of researchers to attempt to bring chips ever closer to the brain's architecture—and in some cases to the architecture of individual neurons. The first electronic neuron was attempted some thirty years ago by a scientist named Leon Harmon, but that effort and others since then have proven to be little more than curiosities, since the functioning of these neurons was rudimentary and there was no practical way to combine them into useful, brain-like circuits. More recently, however, two researchers have developed a "silicon neuron" that can be fashioned together into neural circuits. The researchers, Misha Mahowald of the California Institute of Technology and Rodney Douglas of Oxford University, designed the transistors on the chip to function analogously to the ion channels in the membrane of visual neurons, with the result that the chip appears to duplicate closely many of the observed properties of these neurons. Indeed, some neurobiologists who have examined recordings of the electrical responses of the silicon neuron have had trouble distinguishing them from the responses of real neurons. Terry Sejnowski, one of the researchers who helped popularize neural networks in the 1980s, is convinced the device is the sort of hardware breakthrough for which the field has been waiting. "We're now about twenty-five years away from a silicon brain," he says, beaming.

Some researchers insist, however, that such devices only replicate the most obvious and generic properties of neurons. Until the more complex and subtle neural behaviors are analyzed, they say, silicon neurons will provide at best an incomplete emulation of brain circuitry. Part of the problem with artificial neurons may be, for example, that they are too well behaved and predictable; real neurons are capable of what appears to be almost random behavior, and this seeming "noisiness" may be an integral aspect to their functioning.

But even noise is turning up in the toolkit of artificial neural network researchers. One breakthrough came from Frank Moss, a physics professor at the University of Missouri in St. Louis. While on sabbatical in the early 1980s, the bearded, large-framed Moss stumbled on a paper that described how fluctuations in the outside temperature—that is, thermal noise—helped ensure that the eggs of certain species of turtles would produce equal numbers of males and females. Intrigued, Moss began studying noise, and soon came upon an even more surprising phenomenon, discovered by a group of Italian researchers, called "stochastic resonance," in which noise could actually amplify, rather than interfere with, a small oscillating signal.

Thus when physicist Adi Bulsara at the Naval Ocean Systems Center in San Diego asked Moss in 1990 to consider studying the effects of electronic noise on an artificial neuron, Moss wasn't predisposed to conclude the effects would be negative. In fact, Moss determined that adding the right amount of noise improved the electronic neuron's ability to fire in response to a weak signal by as much as a factor of ten.

Bulsara and Moss assumed this stochastic resonance, while potentially helpful to electronic neurons, was of little relevance to real neurons. But when Bulsara displayed a graph of the effect at a conference on neural networks, one of the attendees recognized it as virtually identical to one that brain researcher John Brugge at the University of Wisconsin had come up with twenty years earlier when studying the firing pattern of neurons in the auditory nerves inside the monkey's ear. As it turned out, an experiment similar to Brugge's was currently being carried out on the vision-processing neurons of cats exposed to a flashing light. That study, by Rutgers researcher Ralph Siegel, kept coming up with exactly the same distinctive graph. Apparently, auditory and visual neurons—and presumably other types of neurons—were enlisting noise to extract more information from the signals they receive. "Usually when you filter out noise the quality of information goes up," explains Moss. "But in this case, if you removed the noise you'd lose an important element of information going to the brain."

Where does neural noise come from? Some would be due to temperature fluctuations and vibrations, notes Moss, but not enough to account for the observed effects. "Noise is ubiquitous in neurons," he says. "Wherever you stick a probe in you get noise. That suggests to me—and this is where some biologists start throwing tomatoes—that evolution has found noise so useful that it is even possible that biological systems have invented ways of enhancing noise."

Moss and Bulsara have continued to work on neural stochastic resonance, investigating its presence in the brain as well as ways of incorporating the effect into artificial neurons. Other researchers, meanwhile, have begun to examine the importance to neural networks of "chaos"—the seeming randomness and unpredictability that emerges in complex systems ranging from weather fronts to dripping faucets. Though chaotic systems are often unstable and prone to wild fluctuations, a mildly chaotic system can be fairly well behaved and yet capable of a wide range of interesting behaviors—just the sort of characteristics one would look for in brain-related phenomena.

Michail Zak at the Jet Propulsion Laboratory, for example, notes that two neurons with chaotic properties might be identical in makeup, environment, and past "experience," and yet would end up producing different firing patterns when exposed to the same signals. Zak contends such quirkiness could be exploited to provide brain-like diversity and responsiveness in neural networks, and has developed neural network models to test his ideas.

Even if researchers come to identify and understand many of the complexities associated with real neurons, and even to figure out how to simulate them—at least in principle—in artificial neural networks, there is no guarantee that such simulations will be readily implementable on a silicon chip with transistorized neurons. Tank wonders if such biologically inspired implementations are even desirable. "One wants to understand the physical properties of neurons and synapses, and how the brain uses them in an effective computing network," he says. "But knowing that doesn't mean you want to build them into silicon. The basic computational primitives for silicon may not be the same as for biochemistry, so maybe you shouldn't try to do things the same way. You have to take into account the basic physics of the device you're using."

Tank's view could be taken as something of a rebuff to those who want to build silicon chips that closely mimic real neurons. On the other hand, it could be taken as encouragement to those who believe that neural network chips should be constructed out of something other than silicon—something, perhaps, that is inherently more neuron-like.

Gen Matsumoto would like a wiring diagram for the human brain. To get it, he is taking movies of the neuron-to-neuron flow of signals within a brain.

Matsumoto is far from the only one employing new imaging techniques to learn about the brain. In the United States and worldwide there has been in recent years a burst of activity in such brain observation technologies as magnetic resonance imaging (MRI), which uses radio waves and magnetic fields to detect changes in the brain's atoms; superconducting quantum interference devices (SQUIDs), which observe minute magnetic fields accompanying neural activity; positron emission tomography (PET), which detects particles emitted by active neurons taking in radioactive glucose; and even optical imaging systems that distinguish changes in the amount of light reflected off parts of the brain that are especially active.

But researchers in other countries are going to find it difficult to match the determination of Matsumoto and other Japanese scientists to peer inside the brain. While New York University researchers boast about their 5-SQUID array, Japanese researchers prepare to bring on-line a 200-SQUID array. As scientists at the University of Washington announce they can detect areas of neural activity in brain patients who have parts of their skull removed, Fujitsu announces it is developing a helmet that will read minds—that is, it will monitor brainwaves and should at least be able to distinguish between a "yes" and a "no."

Matsumoto, for his part, is working with rat brains rather than human brains. The reason is that he is after a more intimate portrait of thought in action than are most other researchers, a portrait that requires dealing with slices of brain taken fresh from its living host. The slices are injected with a voltage-sensitive dye, placed in a dish filled with oxygenated blood, hooked to electrodes, given a mild electric stimulation, and then photographed with a unique camera that zooms in on a few neurons that are each less than one thousandth of an inch wide. As the stimulated neurons fire and pass on their signals to other neurons, the dye glows, providing a visual trail of the path of the signals. The trail is faint, and lasts a fraction of a millisecond (a thousandth of a second), but the camera captures it vividly. From these trails Matsumoto hopes to figure out the paths along which the brain routes signals from component to component. "When I understand it all, I'll build a computer that works the same way," he says.

Specifically, Matsumoto intends to design a neural network based directly on the brain's wiring scheme. Such a neural network, he insists, is not only important to science but to society at large; conventional computers, he claims, are behind many of the world's ills. His reasoning—which is more relevant to Japanese society than to U.S. society—goes something like this: conventional computers can't grade essay questions, so teachers give multiple-choice tests to accommodate computers; to help students score well on these tests, teachers focus on the sort of cut-and-dry, less intellectually challenging material on which these questions tend to be based; because students are raised on this less challenging material, they enter adulthood and professional life unprepared to seek out and deal with difficult problems. Thus society never solves its trickiest problems. Neural networks, says Matsumoto, could end the cycle.

That is only one of Matsumoto's many off-beat theories. Another is that one's sense of self-worth controls everything from mental acuity to physical health; to prove his point, he explains how a pep talk from him instantly cured a young car accident victim's partial paralysis. For what it's worth, Matsumoto's own sense of self-worth is intact, and not without some justification: in addition to being one of Japan's most respected neuroscientists, he looks and handles himself like a thirty-five-year-old tennis instructor rather than the scientist in his fifties that he is.

Matsumoto believes he is close to understanding how the brain is constructed. The key, he says, is not the way the brain performs calculations, which can be easily simulated on a computer, but rather the way the brain stores and recalls memories. To illustrate his theory of memory, he tells a story widely reported in the Japanese press of a wife who asked her husband how his trip to the market went, at which point the husband flew into a prolonged and uncontrollable rage; it later came out (though unfortunately not until divorce court) that when the husband had been a child his father had berated him for returning from the store without the proper change, and the unconscious memory of this incident had been triggered by the wife's innocent question.

This peculiar tale serves to demonstrate the four important characteristics of human memory according to Matsumoto: the deepest memories are of highly stimulating events; memory storage is strongly associated with emotions; memories act at a subconscious level; and strong memories are never erased. To achieve these four characteristics, says Matsumoto, the brain must take the incoming information during a crucial event and shuttle it around to different parts of the brain for various types of processing—such as attaching emotional import—before storing it away; reconstructing the memory requires a similar feat of neural dispatching.

Before developing his thought camera, Matsumoto dissected the brains of thousands of squids (the animal, not the quantum device), whose neurons are extraordinarily large and simply connected. The demands of his research team for squid brains was so large, in fact, that he spent much of two years figuring out how to grow his own squid in a tank, after other researchers had long since given up on the mystery of why the animals always died in captivity. (Matsumoto discovered the squid were dying from ammonium poisoning, and he now keeps a room-sized aquarium brimming with the creatures; raw, debrained squid is a frequent and popular lunch at his lab.)

But trying to determine how a brain directed the signals of millions of neurons by studying individual squid neurons with a probe is like trying to construct an accurate commuting picture for all of greater New York City by running up to cars at stoplights and asking drivers where they are going. The camera was the tool he had been looking for; it made him the first person in the world to see a single thought in action, live and on screen. As he examines the colorful streaks of one neural burst caught with his camera, Matsumoto gestures with his finger across the video image. "We used to think the hippocampus sent signals in this direction," he says, pointing, "but now we know it's a more radial path." Researchers in his lab have already begun work on an artificial neural network that simulates the data flow caught by the camera. In five years or so, he says, he should have enough neural routing data to allow constructing a neural network that simulates the rat brain's complete neural dynamics. Then, imaging equipment allowing, it would be on to humans.

If Matsumoto seems almost wacky in his style and ambitions, it is only because Japanese neural network researchers are far more free-wheeling than their counterparts around the world when it comes to devising schemes to mimic the human brain. Part of the reason for this openmindedness is cultural: on the whole, Japanese are not taught to perceive as strong a dividing line between the natural and the technical as most Americans and Europeans. Thus bridging the gap between a chip and a brain seems a less forbidding challenge to many Japanese researchers than it is to others.

But equally important, Japan has been severely stung by its inability to catch up to the United States in the arena of software. Though it has trounced international competition in many areas of electronics, Japanese companies recognize that conventional computer hardware is increasingly becoming a commodity business of ever-lower profits; the real money in the twenty-first century, most observers predict, will be in software and radically new types of computing devices. In the early 1980s, Japan instituted a government-subsidized project to leapfrog U.S. software technology via expert systems and other conventional AI technologies. This ten-year AI effort, known as the Fifth Generation Project, was no more successful than other conventional AI projects, and for the same reasons: the logical and heuristic approaches to AI are simply inadequate. The project was widely labeled a failure.

But where the failure of conventional AI in the United States in the early and mid-1980s caused the government and industry to pull back somewhat from their support of the field, the Fifth Generation flop only increased Japan's determination to push ahead via AI. Furthermore, Japanese businesses and government agencies, which usually work in concert to fund technical research, are far more encouraging of long-shot endeavors, even if they seem dubious by the standards of U.S. funding agencies. As a result, Japan embarked in 1992 on the Sixth Generation Project, also known as the Real-World Computing

Project, whose stated goal is to achieve human-brain-like computing capabilities by the year 2002. The Japanese government has already committed to kicking in a half billion dollars, and industry could inject billions more. (Computer maker Fujitsu alone spends $2 billion per year on research.)

Interestingly, where the Fifth Generation Project was widely perceived in the United States as a tremendous threat to our lead in computing, the Sixth Generation is not. AI luminary Edward Feigenbaum went so far as to coauthor a 1983 book specifically to warn of the disaster that would befall the United States if it didn't rise to the Fifth Generation challenge. Reviewing the book, *The New York Times* wrote of "Japan's commitment to produce within a decade a new generation of computers so immensely powerful that they will in effect constitute a new and revolutionary form of wealth." Here, on the other hand, is what *The New York Times* reported in its only article of 1992 on the newly formed Sixth Generation Project:

> For the most part it is eliciting more puzzlement than fear [in the United States]. Indeed, some computer scientists say the new project shows that [Japan's] vaunted system of Government-led industrial development is running out of steam, at least in electronics, just at a time when other nations are thinking of emulating Japan.

What's more, the U.S. government loudly complained ten years ago that the Fifth Generation Project was excluding U.S. researchers; the formation in the mid-1980s of MCC and Lenat's Cyc project was a product of the government's concern. But this time around the government has actually encouraged researchers to *turn down* Japan's enthusiastic invitations to participate in the Sixth Generation, claiming Japanese laws do not offer adequate protection over intellectual property.

Why so blasé this time around? Part of the reason may be that conventional AI researchers are still the big guns when it comes to influencing research, and many of them are no more interested in seeing money poured into neural networks now than they were to see it poured into perceptrons in the 1960s and 1970s. But in addition, even among many U.S. neural network researchers there is a hint of derision toward the Japanese enthusiasm for achieving brain-like devices. The fact is, when Japanese researchers gush about constructing human brains, they sound a little silly. On the other hand, they probably don't sound a lot sillier than the Japanese automakers did to their U.S. counterparts in the early 1970s when they emphasized practical, high-quality, fuel-efficient cars.

Shun-ichi Amari, a highly respected University of Tokyo researcher and chairman of the planning committee for the Sixth Generation Project, is sitting in a lounge of the New Otani hotel just before delivering the opening speech to the first official meeting of the undertaking. Amari seems anything but naive about the task ahead. "The human brain is the most complex thing mankind has ever tried to study," he says. "Duplicating it may be impossible. But we think that with enough effort, we can put together pieces of the picture that will open up new generations in computing."

Some of the pieces undertaken by Japanese scientists seem like solid neural network research embellished with a fanciful twist. Mitsuo Kawato at the Advanced Telephony Research Labs (Japan's Bell Labs), for example, states up front that he wants to build a robotic "secretary" within ten years. "It will have a fax and a telephone, and it will know how to keep out of your way," he explains. Kawato's robosecretary will also know how to speak naturally, if his research pays off; he and associates have taken detailed measurements of the velocity and acceleration of twelve different muscles around the human mouth and throat as they articulate various sounds, and have applied these data to a neural-network-controlled audio system.

And even if the robosecretary doesn't get out of the way fast enough, it shouldn't cause any serious problems: Kawato's group has developed a neural-network-based robot arm that, in imitation of the human arm, incorporates two opposing sets of hydraulic "muscles." As a result, the robot arm is as strong as ordinary industrial robots, but when a person walks into its path the arm bounces off harmlessly as the opposing muscles restrain it. (Like most leading Japanese neural network researchers, and unlike most of his U.S. counterparts, Kawato's networks are highly feedback-oriented. In fact, the feedback-oriented

Stephen Grossberg is considerably more popular in Japan than he is in the United States, which is slightly reminiscent of the way U.S. "quality" guru Edward Demming was revered by Japanese businesses while remaining virtually unknown to American businesses.)

Where the gleaming and even lavish facilities of the ATR Labs is typical of first-rate industrial research facilities in Japan, the cramped, chilly, and slightly dingy laboratories at Osaka University are fairly typical of academic research facilities there; in some university labs, the only thing that gleams is the oddly near-ubiquitous nude pin-up calendar on the wall. But there is no pin-up in the straightlaced Kunihiko Fukushima's lab, who, if nothing else, seems much too busy for such distractions. Fukushima has developed an array of ART-like neural networks that tackle a number of different tasks, including character and speech recognition.

Character recognition in Japanese is a qualitatively more difficult chore than it is in English: there are thousands of characters—most representing complete words—and each is composed of various combinations of some three hundred different graphic symbols. Fukushima's network can distinguish perhaps a tenth of the major characters so far; it can even differentiate between such similar characters as those for "tree" and "small forest," even though they are identical except for a subtle shift in the position of one of the strokes.

To improve his speech recognition network, Fukushima is studying data from research on the neural firings in living monkeys exposed to various noises. But, typically for a Japanese neural network researcher, Fukushima professes grander ambitions. "I'm concentrating on pattern recognition right now because we have enough data on it," he says. "But what I really want to do is build a neural network that represents the entire human brain."

More modestly, Kazuyuki Aihara at Tokyo Denki University merely wants his neural network to duplicate human memory. To do that, Aihara, a wiry, lively young man with a Beatle haircut, has enlisted chaos. By "tuning" the ten nodes in his hand-wired neural network box (he says he has a design for a chip version), Aihara is able to bring his network through various chaotic states that exhibit different memory storage and retrieval properties—some of which, he claims, make his network more human-like than ordinary networks. For example, when presented with an input pattern on which it has been trained, his chaotic network won't always immediately produce the expected output; instead, it will unpredictably "wander around" through a number of various alternative outputs, sometimes suddenly seizing on one of these alternatives as a better answer. That behavior, he says, evokes the human mind's occasional tendency to bounce off in a new direction and come up with an unexpected memory or idea, seemingly out of nowhere. "It's not random," notes Aihara. "The chaotic dynamics of the network provide a kind of grammar for memory."

Given Japanese scientists' unrestrained enthusiasm for unusual neural network designs, it is not surprising that many researchers there have thrown themselves into the question of the appropriate hardware implementation for neural networks. Neural network chips are already entering the mainstream there: Ricoh, for example, has placed a sixteen-neuron, sixteen-synapses-per-neuron chip into a line of copying machines. The company claims it plans to "stack" the chip to form massive, trainable networks that could be incorporated into factory robots.

But more than in the United States, there is wide recognition in Japan that conventional implementations of artificial neurons are probably inadequate, propelling a number of researchers there to look for alternatives to silicon. At Matsushita, Katsuhiro Nichogi has been experimenting with various combinations of metals and semiconductors, and has come up with an aluminum-germanium device that provides a close mimic of neurons' nonlinear and adaptive responses; because these responses are a natural property of the substrate, and not of the complex circuitry of "silicon neurons," artificial neurons made of this substrate could theoretically be made far smaller, faster, and neuron-like than other versions. Masahiro Okamoto at the Kyushu Institute of Technology is taking an even more biological approach: he has fashioned a set of neuron-like switches based on a suite of chemicals known as a "cyclic enzyme system," a system employed in living organisms to regulate metabolic processes.

But Japanese researchers are not about to overlook the most neuron-like option of all: living neurons. A few years ago, placing live neurons onto a chip was not an idea that many scientists would have been willing to run with, due to some formidable obstacles—such as the fact that no one knew how to grow them, hook them together, or conveniently get signals in and out of them on such an unconventional medium. Individually, these were tough problems; together, they appeared overwhelming. Which, at least as far as one researcher was concerned, was a pretty good reason in itself to attempt it.

Growing up in Yokohama in the late 1940s, Masuo Aizawa preferred studying history to science. But history seemed an almost frivolous endeavor in the context of the pressing realities of postwar Japan, and so when Aizawa entered Yokohama National University he reluctantly majored in chemistry. In his third year a lecture by visiting Tokyo Institute of Technology Professor Jun Mizumuchi ignited Aizawa and set the direction of his career. Mizumuchi made the then-outlandish prediction that biology was going to have a huge impact on a wide variety of technologies in the coming years. Captivated by the possibilities of applying natural processes to artificial devices, Aizawa decided on the spot to throw himself into the mechanics of biological systems.

By the time Aizawa became a graduate student at the Tokyo Institute of Technology in the 1960s, he had created his own specialized field of study—bioelectrochemistry—to fulfill Mizumuchi's prediction. His first breakthrough was a "biobattery" constructed out of electrodes inserted in a jar filled with cellular proteins called mitochondria, which extract energy from sugar and store it as electric charge. The biobattery's feeble 1.5 volts, as well as the tendency of the complex proteins to break down, precluded its application as a commercial battery. Unfazed, Aizawa converted his biobattery into a supersensitive glucose detector: when even trace quantities of glucose were present, the device put out a tiny but detectable current. It was eventually developed into a version that can, among other applications, help diabetics monitor their blood sugar level. Aizawa went on to design the first "immunosensor"—a device that employs antibodies of the sort found in our immune system to ferret out and lock onto almost any sort of molecule. Immunosensors (Aizawa coined the name) have since become a hot technology worldwide.

Around 1980, Aizawa began to wonder if there weren't some way to harness the most complex cell of all: the neuron. If he could somehow couple neural cells to an electronic device, it occurred to him, it might have some applicability to artificial neural networks, which were at that time starting to regain attention among computer researchers. But that would require growing neural cells on electrodes—that is, on some sort of conductive surface—so that electrical signals could be inserted into and extracted from the cells. In 1980, that was an outrageously farfetched idea; even ordinary animal cells hadn't been grown on electrodes, and neural cells are so much more delicate that it was all but impossible at the time to culture them in even the most hospitable media.

Now at the University of Tsukuba, Japan's famous "Science City," Aizawa decided to develop the necessary techniques on ordinary cells before applying them to neurons. He tried to get various cells to grow on a number of different conductive and semiconductive materials, including gold, platinum, titanium, carbon, and conductive plastics. The best results, as it turned out, came with the semiconducting compound indium tin oxide (ITO), but even then the cells grew sluggishly and unevenly. On a hunch, Aizawa tried placing a small voltage—on the order of a tenth of a volt—across the coating of the ITO; he reasoned that because the surface of a cell maintains its own small voltage, the cell might be receptive to a similar voltage in its surrounding medium. Sure enough, the tiny voltage stimulated cell growth, causing the cells to proliferate via division across the entire slide.

But Aizawa knew that even if the charged ITO environment proved friendly enough to grow neural cells, he would still have a major problem. Neural cells grow by sending out long, tentacle-like axons and dendrites, generically called "neurites." It is, of course, through the dense, interconnected webs of neurites that neurons transfer electrical signals to one another. But if the neural cells growing on his slide were to freely throw out neurites in every direction, he would end up with a dense sprawl of haphazard growth that would defy any efforts to study, let alone influence, signal transmission.

To build a primitive neural network, Aizawa knew he would need an orderly array of neural cells; in fact, the best way to examine signal transmission would be with a long, single-file string of connected neural cells. That, realized Aizawa, would be a useful first step in trying to design a semiartificial neural network. With a string of neural cells, he reasoned, it should be somewhat easier to introduce a voltage at one end and then detect the output signal at the other end of the string, or anywhere in between. It would also allow trying to perfect techniques for strengthening various neural connections through repeated firing, and perhaps to discover other ways of influencing the transmission of signals. Once the properties of neural strings were mastered, the strings could be fashioned into a checkerboard pattern—a pattern that might be induced to function something like a simple, one-layer perceptron. But how could neural growth be directed to produce such strings?

Once again, Aizawa turned to electricity as a means of controlling biology. He continued to study ordinary animal cells, exposing the cells growing on the ITO to a wide variety of voltages. He soon discovered that voltages above 0.5 volts usually proved fatal to the cells, while those below 0.3 tended to stimulate their growth or have no effect. But a voltage of about 0.4 volts had a completely unexpected effect that had never before been observed: it stopped the animal cells from dividing without otherwise affecting their function in any way. "I was amazed," says Aizawa. "It was as if they went into hibernation." He realized that discovery could be exactly the one he needed: if the right voltage froze animal cell division, perhaps it could also be employed to control neurite growth.

In 1985 Aizawa returned to the Tokyo Institute of Technology to found a school of bioengineering—a facility that soon had over 100 researchers and technicians. By 1987 he had resumed his research, but this time he was ready to try his hand at neural cells. To improve the odds, Aizawa and graduate student Naoko Motohashi (one of Japan's relatively rare female scientists) decided to work with a type of cell known as PC12 rather than jumping into brain cells. Derived from connective tissue in rats, PC12 starts off as an ordinary, dividing, round animal cell, and stays that way until it comes into contact with a substance known as nerve growth factor, or NGF. Then PC12 immediately stops dividing and within three days starts to grow neurites. Within two weeks the PC12 cell is transformed into a fully functional neural cell. Aizawa hoped that because it starts off as an animal cell, PC12 would prove hardier and less fussy than neurons.

At first, it didn't; the PC12 cells wouldn't reliably grow into neural cells on the ITO. But Aizawa and Motohashi kept at it, varying the voltage, the temperature, the thickness of the ITO, the ingredients of the fluid in the petri dish in which the slide was submerged (the main ingredient was calf plasma). After several months, they finally had neural cells growing on the ITO—but the cells didn't always respond to their efforts to freeze neurite growth with a higher voltage. For more than another year the researchers experimented with voltage, varying the strength and the timing of the applied charge. Sometimes a certain voltage worked, and then other times it wouldn't. "After a while, we started to have doubts about whether this phenomenon could be made reproducible," recalls Aizawa.

Finally, in 1989, the two scientists were able to produce slide after slide of PC12 neural cells arrayed in alternating stripes: the cell-less stripes corresponding to those bands of ITO that had been laced with 0.4 volts, while the other, lower-voltage bands of ITO boasted dense growths of interconnected neural cells. The neurites of these cells crowded along the edges of the higher-voltage electrodes but didn't cross over onto them. The reason the electrified areas remain free of neurites, speculates Aizawa, is that 0.4 volts may be just enough voltage to realign charged molecules on the surface of the cell into a shape that blocks the entry of NGF, preventing neurite growth. Even if a neural cell starts off on the electrified stripe, it migrates off it within a few days. "I don't know how it does that," says Aizawa. "I think maybe it rolls."

Aizawa has continued to work on refining his control over the neural cell growth. He has now achieved a rough version of the sought-after neural strings, stripes of interconnected cells a mere thousandth of an inch wide. The next step is to design an input and output to his string: that is, a way to introduce artificial electronic signals into the string and to detect the resulting signals that emerge from the other end. Normally, researchers insert signals into neurons by sticking a probe in them, a technique that ultimately kills the cell. Aizawa wanted a noninvasive, nondestructive method.

Fortunately, he is already halfway there by virtue of having grown the cells on top of a semiconductor. He is now trying to develop a grid of electrodes so he can selectively stimulate the individual neural cells on top of it. With luck, the same electrodes could be used to extract the resulting signals from other neural cells. If he can perfect that technique within a few years, as he believes he can, Aizawa will start to investigate ways of using signals to strengthen the connections between neurites, a prerequisite to neural programming. If that works, he could then attempt to fashion a simple, programmable, single-layer neural network. He predicts it will be ten years before he gets that far—"If I can do it all," he adds.

If his preliminary neural chip proves able to perform rudimentary tasks such as recognizing simple patterns, the next step will be to try to build a three-dimensional structure of neural cells that form a multilayer neural network capable of more complex functions. That would open up any number of possible applications. Aizawa notes, for example, the possibility that a neural chip could be used as an interface between prostheses and the nervous system for patients who have lost limbs. (Stanford researcher Bernard Widrow and physician Greg Kovacs are already working on techniques for attaching the nerve endings of severed hands to a chip.) Aizawa's chip could perhaps even be connected at one end to a tiny video camera and at the other to the optic nerve, creating an artificial eye.

At a minimum, Aizawa's chips should provide invaluable insights into how neurons process signals, a mystery that continues to limit the effectiveness of artificial neural networks. But, of course, Aizawa holds bigger hopes for his chip. "We want to be able to build at least a part of a brain," he says.

Since having to connect up individually millions or even billions of neurons—real or artificial—is a hopeless task, the challenge to constructing even part of a brain is to find a way to harness the phenomenon of self-organization, as nature does. But neurons sitting in a dish do not, as Tomaso Poggio might say, magically organize themselves into a useful neural network. Even in the brain, neurons are not discrete, self-propelled nodes that move themselves into the proper position. Rather, neurons self-organize through biological processes that take place on a far smaller scale than the neurons themselves. Accomplishing the same in an artificial context, then, may very well require mastering these processes, or others like them. In other words, nature-based AI is increasingly finding itself pushed into the realm of the submicroscopic.

5. MOLECULAR INTELLIGENCE

Am I supposed to be doing this?
—STEEN RASMUSSEN

I t's time to feed the Metabolism.

Peering into the clear liquid in the beaker, Steen Rasmussen sprinkles in white powder from a small bottle. The powder quickly melts into delicate milky clouds that lazily plume and roil before dissolving into transparency. But though there is little for the eye to see, the Metabolism is busy. It has begun to feed on the powder, and to grow. It is hardy, adapting to changes in temperature and acidity. But if deprived of food, or exposed to too severe a change, the Metabolism will . . . well, die.

Actually, the Metabolism is a deceptively simple concoction of biomolecules—they happen to be strings of DNA, but in theory they could be other kinds of molecules—that are constantly joining together to form longer molecules. What's special about this jumble of activity is that each of the many chemical reactions going on lends a hand to one of the others; once the chemicals are provided, the reactions begin to reinforce each other until an ongoing, self-maintaining symphony of chemical interactions has pulled itself up by its bootstraps. As long as it is provided with the chemicals it needs to kick off the initial reactions, this "cooperative chemical network" can sustain its activity for long periods of time, requiring only occasional "feedings"—that is, replacing the one chemical that happens to break down quickly.

Rasmussen's odd brew represents a crude form of self-organization. Nevertheless, it represents a first step toward providing a critical missing ingredient to artificial intelligence, an ingredient whose absence has led a growing number of researchers to explore realms so far below the size scale of ordinary lab work that just a few years ago such efforts would have been completely unthinkable. These scientists are now pursuing individual molecules.

Why go so small? For one thing, life itself arises at the biomolecular level; and as the principles of nature-based AI dictate, questions of intelligence are impossible to separate from questions of life. Whatever it is that molecules are doing to create living organisms, chances are their activities play a crucial role in imbuing these organisms with intelligence. To replicate intelligence, researchers will have to learn to play the same game.

It's not just nature-based AI researchers who want to go to smaller scales. Any form of AI, and for that matter any computing task, would be improved by faster hardware. And when it comes to faster hardware, small is the name of the game.

The phenomenal improvements achieved in the past few decades in the speed and cost of computers and electronics have been financed by the continuing ability to make ever-smaller circuits in which electrons have shorter distances to travel. Now the great hunt is on for the next generation of technology that will carry circuitry down below the size scale of today's silicon transistors. Unfortunately, the semiconductor industry is approaching a roadblock of sorts: transistors are shrinking below the millionth of an inch mark, the size scale at which individual atoms start to stand out.

The problem is that when any structure gets below a millionth of an inch, the rules and equations that cram engineering textbooks become obsolete, and quantum mechanics steps in to take their place. This is the domain of the uncertainty principle, where particles are wave-like entities that are here one second and over there the next without ever having been anywhere in between. Right now, the road to further miniaturization is looking increasingly ravaged by quantum mechanical potholes. A transistor, which acts as a sort of microscopic traffic light at the busy intersections of a chip's millions of electronic roadways, isn't very effective when the electrons are constantly disappearing and popping into existence further down the road.

The need to develop electronic components smaller than transistors without butting heads with quantum mechanics is the motivation behind the Center for Quantized Electronic Structures, or QUEST, a sort of quantum engineering think tank founded in 1989 at the University of California in Santa Barbara. Consisting of about twenty physicists, chemists, electrical engineers, and materials scientists, many of them raided from Bell Labs, QUEST has been favored with a $25 million annual grant courtesy of the United States government— which, all things being equal, would just as soon that the dominance of the next generation of microelectronics be achieved by the United States and not Japan.

The scientists and engineers at QUEST are the leading practitioners of what might be called the hot-dog school of microelectronics. Their philosophy: if quantum mechanical waves are rocking the boat, then learn how to surf. By assembling atoms into structures that nature never thought of—structures mere billionths of an inch in scale—these researchers are forcing a kinder, gentler behavior out of quantum mechanics that actually works in their favor. "Other people set limits on how small you can go because of quantum mechanical effects," says QUEST director James Merz. "We're finding ways to harness those effects."

If QUEST's efforts at atomic bricklaying are new, the basic theory behind it has been bandied about for nearly seventy years. The framework for quantum mechanics had barely been laid shortly after the turn of the century before theorists were predicting some interesting properties for electrons confined in a small space. That's because quantum mechanics describes an electron in a box as a wave, complete with crests and troughs, spread throughout the box's volume. Just as a basketball player could add energy to a basketball by dribbling more rapidly—squeezing in, say, twenty-five bounces instead of fifteen as he traverses the length of the court—so a boxed electron becomes more energetic when extra crests and troughs are jammed into the same space, shortening the electron's "wavelength."

The catch is that the electron can only pick up an additional trough and crest as a complete set; after all, it wouldn't do a basketball player much good to leave the ball on the floor as he took off for the basket. That requirement isn't a big deal in a box much larger than the billionths of an inch length of an electron's trough and crest, because sneaking an extra trough and crest into all that space would barely change the shape of the wave. As a result, such an electron can smoothly add to its energy in tiny increments. But in a box that is barely wide enough to hold a single, low-energy trough and crest to begin with, adding an extra trough and crest would require scrunching up the wavelength in half—and thus doubling the electron's energy. In this way, a severely confined electron is forced to make its way up the energy ladder by leaps and bounds instead of smoothly. Or, in physics jargon, the electron's energy levels become "highly quantized." In fact, that's what happens in an atom: because an electron is bound in a billionths of an inch orbit around the nucleus, the energy associated with its motion becomes quantized, limiting the electron to certain orbits.

In the mid-1970s, the notion of fitting electrons with quantum mechanical straitjackets began to seem less farfetched. By the end of that decade, a team of researchers at Bell Laboratories had constructed a wafer comprising three layers of semiconductors—materials like silicon, gallium arsenide, and aluminum gallium arsenide that allow electrons to move freely within them only when subjected to a small voltage—that prevented electrons in the middle layer from moving up or down toward the other two layers. The trick to producing this "quantum well," as it was dubbed, is to vary the ingredients of the middle layer so that electrons can move within it a little more easily than they can in the outer layers; thus most electrons in the middle layer find themselves unable to muster up the energy to jump into the more restrictive upper or lower layers. In addition, if the middle layer is made extraordinarily thin—just a few layers of atoms—the electrons trapped in this layer have no headroom. The result: like marbles rolling around a tabletop, the electrons are restricted to moving in a plane.

Even though the electrons in a quantum well are still free to careen through the middle layer, the fact that they have lost one dimension of motion is enough to cause quantum mechanics to rear its curious head, and the first glimmers of energy quantization start to show up. Like a double-teamed basketball player who finds many of his options cut off, an electron hemmed in by a quantum well has fewer ways to alter its wavelengths, and thus some energy states start to become less accessible to it. That sort of property is invaluable in the design of lasers, for example, which depend on the light energy emitted by electrons dropping from one particular energy state to a lower one. In fact, the laser beams in most compact disk players are produced by chips based on quantum wells.

Quantum wells are just the jumping-off point for QUEST. Researchers there want to limit electron motion to only one, or even zero dimensions, by constructing quantum "wires" and "boxes," respectively. In a quantum wire, electrons are a little like racehorses with blinders on: they can't be slowed by interacting with atoms that are above, below, or to either side of them, because they can only move straight ahead or backward; as a result, they move "ballistically," or nearly without resistance, at up to 100 times the speed of electrons on ordinary chips.

In a quantum box, a sort of tiger cage for electrons, they don't move at all. That may not sound very useful, but it means that their energies become so perfectly quantized that researchers drool over the possibilities. "We're really talking about artificial atoms, about constructing matter with entirely new properties," says Merz. "We can't even guess at some of them, but we know we should be able to assemble quantum boxes into materials that have whatever conductive properties we need." Quantum boxes would also have extraordinary "optical" properties—that is, the way in which they emit, reflect, and absorb light. An electron's inclination to interact with different frequencies, or colors, of light depends on its energy level. Because the energy levels of a quantum-boxed electron are widely spaced, a single energy jump can profoundly alter these interactions. Thus a small electric jolt might instantly transform an array of quantum boxes from dark to brightly glowing, red to blue, or mirrored to transparent.

To build a quantum wire, QUEST's scientists employ "molecular beam epitaxy" devices, exotic machines that boil off molecules from a chunk of semiconductor and send them hurtling through an ultra-high vacuum toward a semiconductor wafer in a target chamber. When the rain of molecules hits the wafer they stick via their electrically charged components, retaining enough freedom of motion to avoid clumping and thus settling into a flat, ultrathin layer. By alternating layers, they can produce a stack of quantum wells, or "superlattice." By offsetting and varying the thickness of each layer, furthermore, they end up with a snake-like distortion in the tilted superlattice—and, like a pig in a python, electrons are trapped at the bulges and are confined to moving in a straight line along them, quantum-wire style. These "serpentine tilted superlattices" are the state of the art in quantum electronic devices; 10 million of these quantum wires would fit inside a human hair, and QUEST can fabricate them at a rate of trillions per hour. QUEST scientists hope to soon use a beam of atoms to punch indentations into their quantum wires, so as to close off sections and create quantum boxes.

Achieving the phenomenal computing speeds promised by chips based on quantum wires and dots would provide a huge boost to conventional AI applications, some of which function reasonably well in a limited domain but are simply much too slow. Processing speed breakthroughs would also make a world of difference to neural network researchers, who could build larger, more complex, and better performing software simulations.

But for nature-based AI, the main motivation for moving to the molecular scale is not merely to speed up existing approaches, but to be able to create new approaches using the same building blocks that nature does. If nature-based AI is to take a truly bottom-up approach, then it needs to look at the very bottom: biomolecules.

Fortunately, AI researchers interested in harnessing biomolecules can build on the work of a number of researchers who have been intent on smudging the line between the manufactured and the natural. Their simple credo: Why reinvent the wheel? Nature, after all, has spent billions of years perfecting a dazzling orchestra of hard-working biomolecules for sustaining life. "Biomolecular systems have such fantastic properties in and of themselves, and they literally grow in trees," says Stanford chemist Steven Boxer. "We've decided that since we can't beat them, we should join them."

Not that biomolecules are easy to harness. They are so small, so complex, and in many cases so prone to malfunctioning outside their ordinary venues that working with them can be a study in frustration. But undaunted by the formidable challenges, scientists are attacking the obstacles with a mélange of tools and techniques culled from several disciplines. Genetic engineering makes it possible to tailor a protein's makeup, and hence its function. Organic chemistry provides new materials for anchoring and preserving the altered proteins. And electrical engineering offers ways to detect signals from the internal workings of the proteins. When it all comes together, the result is the conversion of a small assembly of biomolecules, or even an individual biomolecule, into a custom-designed machine.

One example comes from Stephen Mann at the University of Bath, who with his colleagues is turning proteins into submicroscopic chemical processing plants. The group has been focusing on ferritin, a protein found in the livers of humans and a wide range of other organisms that forms eight-nanometer-wide cages (a nanometer is a billionth of a meter, or a few hundred millionths of an inch) with a strong affinity for iron oxide. When free iron in our bodies picks up oxygen and forms rust, it does so inside the ferritin structures, which keep the toxic compound safely caged. "We had been studying the structure and properties of the native protein," recalls Mann, "when it occurred to us that we might be able to use the protein as a reaction vessel for controlling the particle size of other materials."

Mann and his team found that the protein can cage several other compounds besides iron oxide, including manganese oxide and iron sulfide. The next step is to alter the chemical affinities of the protein itself to get it to trap still other compounds. "We've identified the two key sites in the protein involved in the specific mineralization product," says Mann. "In principle, we should be able to engineer the protein to fit the mineralization product we want inside it." That possibility might appeal to semiconductor researchers, who are now striving to form minute structures with unique quantum-mechanical properties. What's more, caged particles would enjoy an unrestricted ride through the human body, where they might be useful for diagnosing or treating disease.

University of Utah biologist David Blair has been studying the twenty-five-nanometer-wide molecular motor that powers the propeller-like flagellum of many bacteria. Spinning at up to 18,000 rpm, the motor pushes an average-sized cell 30,000 nanometers, or about fifteen body-lengths, each second—and it's reversible, too. "It's a triumph of engineering," gushes Blair. To tease out the motor's "parts list," Blair has been altering genes coding for proteins in the motor and studying the effect of each change. So far, the only part he's clearly identified is the "fuel injector"—a proton channel that provides the motor's energy source—but he expects to be able to nail down parts corresponding to a rotor, stator, motor mount, and transmission.

Blair notes that transferring the motor intact to a different structure would be difficult, but he doesn't rule it out. He also points out that the motor seems to be constructed of molecular rings that might be useful by themselves as Tinkertoy-style connectors or as "junction boxes" in other structures. "Or maybe we can use the motor to make really tiny CD players," he says. "Just kidding."

To install the motor in a larger structure, a bionanoengineer might require a nanoscale scaffolding. And for that purpose, New York University biochemist Nadrian Seeman thinks DNA may be just the thing. Seeman credits his original idea that DNA could be coaxed into a lattice structure to an M. C. Escher print. Last year, by altering key sequences in the molecule, he got an assembly of DNA branches to fold itself into a cube, and he's confident more complex structures will soon follow. "The next step is to have other molecules associate with the DNA," he says. "Then you could have molecules with electronic properties riding the DNA into place, forming circuits."

Other researchers picture biomolecules in more active roles. Some hope, for example, to capitalize on biomolecules' extraordinary sensitivity to their environment in order to make improved biosensors. Conventional versions of these sensors, widely used in medical diagnosis, typically rely on detecting chemical changes triggered by enzymes that have locked onto a target molecule. But in the new, more sensitive biosensors researchers have in mind, the signal would flow directly from the internal workings of individual biomolecules.

Biosensor developer Felix Hong of Wayne State University Medical School, for example, is exploiting a light-sensitive protein called bacteriorhodopsin, a substance found in the purplish membrane of the bacterium *Halobacterium halobium*. This bacterium hangs out in the waters of coastal marshes where the concentration of salt is some six times that of normal seawater, enough to squelch most other forms of underwater life. (The purple splotches of terrain visible from the air around San Francisco are actually salt marshes loaded with *Halobacterium halobium* and its colorful membrane.) In *Halobacterium halobium*, bacteriorhodopsin kicks off the process of photosynthesis, the conversion of light energy to chemical energy. Bacteriorhodopsin's cousin, rhodopsin, is found in mammals' retinas, where it is used to trigger neurons at the back of the eye, ultimately providing us with vision.

Like all proteins, bacteriorhodopsin consists of complex Tinkertoy-like arrangements of the carbon, nitrogen, oxygen, and hydrogen atoms that serve as key building blocks to all living organisms. When exposed to various forms of energy, proteins undergo a "conformational" change—that is, they change their shape. In the process of refolding themselves they generally change their properties, too. Hong takes advantage of the fact that exposure to light causes a conformational change in bacteriorhodopsin that results in its pumping out a proton, whose sudden presence creates a minute electronic signal that Hong can detect. The pumping action, and hence the signal, depends on the molecule's chemical environment—and the specific chemical sensitivity changes with the frequency of the light. Hong has discovered, for example, that when the molecule is exposed to one frequency, it emits a signal that falls off as the acidity of the medium increases, while a different frequency leaves the molecule sensitive only to the concentration of chloride ions. The result: a prototype dual-function biosensor that, in response to a simple shift in light frequency, can switch between measuring acidity and chloride concentration.

Biomolecules could be applied to electronics as well, perhaps even serving directly on chips in place of conventional transistors. Stanford's Boxer, for example, is pinning such hopes on a bacterial protein that responds to light, in this case by pumping out an electron.

In nature, the electron displaced by this "photosynthetic reaction center" drives chemical reactions that provide energy for the organism's vital processes. Boxer instead gets the reaction center to hand off the electron to other proteins or directly to metal electrodes. Because the light-driven handoff can be switched off and on with a second beam of light or with an external electric field, the protein could theoretically serve as a transistor-like element in an electronic or optoelectronic component.

Right now, Boxer is wrestling with the task of getting the protein to stick to electrodes and other surfaces, a feat he accomplishes by genetically altering the bacteria so they produce a modified protein equipped with small molecular hooks. "Once you build the interface between biological molecules and electric circuits," he explains, "there are a lot of things you can do."

Why bother making protein transistors, when researchers like those at QUEST are busily shrinking microcircuits made of conventional semiconductors? It's because the reaction center is orders of magnitude smaller than even quantum wire circuits; what's more, Boxer sees the promise that researchers might someday find a way to coax the components to connect themselves up right in the vat. "The real advantage to making things out of biological molecules is that they can self-assemble," says Boxer.

There may be an even better way to employ biomolecules than to use them as replacements for transistors on electronic chips. One researcher has found a way to produce computing components that do away with electronics altogether, relying solely on the unique properties of biomolecules.

Buried within the bed of protein in bacteriorhodopsin is a molecule called a chromophore, a chain-like construction of carbon, nitrogen, and hydrogen. The chromophore has a unique property: when just the right amount of energy is added to it, a section of it twists, or "flips," in much the same way that a carefully measured kick to a long box lying on its side can knock it onto its end. And just as the box can be knocked back onto its side with another, less powerful, blow, the chromophore can be untwisted by subjecting it to a second, weaker, burst of energy. Different color light beams have different energies, and it so happens that green light has exactly the right energy to cause the chromophore to flip, and red light carries just the punch needed to unflip it.

Robert Birge at the University of Syracuse has put the chromophore's flip to work for him in an entirely novel type of computer memory. A layer of bacteriorhodopsin just a few hundred thousandths of an inch thick is spread out on a mirrored surface and kept cool in a bath of liquid nitrogen, which speeds up the flip. Recording information on the bacteriorhodopsin is simply a matter of socking those chromophores destined to be 1's with a green laser light, causing them to flip, and leaving those earmarked as 0's unflipped. To "read" the information, the device scans the bacteriorhodopsin with a red laser light. When the red light hits a flipped, or 1, chromophore, the chromophore absorbs the light and unflips (a touch of green light mixed in with the red immediately flips it back again). A 0 chromophore, on the other hand, ignores the light and lets it pass through to the mirrored surface, which reflects the light back to a photodetector lurking above. Thus if the photodetector sees a red flash, the device knows the chromophore below it is a 0; otherwise, the device assumes it's a 1.

To help meet his goals, Birge—a Phil Donahue ringer who is also a composer whose works have been recorded on some twenty albums— has convinced the University of Syracuse to pull together the Center for Molecular Electronics, of which he is the director. The Center has a $2.5 million annual research budget, along with 20,000 square feet of first-class research space in a brand-new building paid for by the state of New York, which hopes to turn the area around Syracuse into the molecular version of Silicon Valley. The Center's staff includes eight chemists, two physicists, two electrical engineers, and a biochemist, along with twenty-three graduate students in various disciplines. And while most molecular computing projects are still struggling through basic research, a California company called Biological Components Corporation is already seeking investors to turn Birge's prototype for a computer memory into a merchantable product. "I don't know whether it will be twenty years or thirty years," he says, "but there's no question in my mind this technology will dominate computing." Birge's biomolecular memories are rumored to already be standard equipment on some jets flown by the Navy, which apparently likes the fact that the heat generated in a crash would instantly wipe out the memory and keep information from falling into the wrong hands.

Birge is also working on a vaguely neural-network-like memory that stores information about a picture as a hologram. A hologram takes advantage of the fact that light is a wave, and that when two beams of light are combined the crests and troughs of each of their waves will sometimes reinforce each other and sometimes cancel each other out, resulting in a complex pattern—just as dropping two stones side by side into a pond produces two sets of circular ripples that combine in intricate ways. In the case of the light beams, the resulting pattern takes the form of a series of alternating light and dark areas called an interference pattern. To create a holographic image of an object, a laser beam is bounced off the object, split in two, and then the two beams are recombined to create an interference pattern; that pattern is what's stored on the film. Shining a laser beam through the holographic film reforms the image of the original object. In Birge's system the holograms are recorded on a thin layer of bacteriorhodopsin with a green laser.

In operation, the memory puts the unidentified image into holographic form, projects it onto the bacteriorhodopsin film containing all the reference images, and picks out the image that lines up the most closely. The technique wouldn't work well with ordinary photographs, because even photographs of the same object are rarely similar enough to closely line up. Holographic interference patterns, on the other hand, line up pretty well as long as the two images bear a strong resemblance to one another.

Birge expects eventually to get his memory to record up to tens of thousands of images on a single bacteriorhodopsin film, more than enough to handle applications like handwriting identification and map recognition, and eventually in robot vision and other areas of artificial intelligence. "This is the one area where we can completely blow away our semiconductor competition in six months," he says. "We're going to be able to have the equivalent of 20 million characters of associative memory on a single film that you can instantly read or write on. You simply couldn't build a semiconductor associative memory that large."

A third Birge project is a biomolecular version of a logic gate, the basic building block of conventional computational circuits. The gate brings together three chromophores into a Y shape. One of each of their ends is chemically bonded to a fourth molecule at the center of the Y called a porphyrin, a substance that carries oxygen in blood. As laser light strikes a chromophore in the Y, causing it to flip, the flip twists the attached pie-shaped porphyrin around like a sort of lazy susan; the porphyrin's contortion, in turn, twists another chromophore, causing it to flip. Birge's short-term goal is to employ the gate to construct a sort of molecular calculator capable of adding two small numbers. Not exactly an earthshaking feat of computation, but the device will be able to come up with its answer in less than three trillionths of a second, a thousand times faster than a supercomputer could do it. There is still at least one big problem: the gates tend to fall apart after about six months, as the relatively weak electric bonds that hold them together succumb to the pull of neighboring molecules.

Of course, biomolecules are subject to the same quantum mechanical limits that semiconductors face. In fact, because they are only billionths of an inch wide, biomolecules already suffer from the erratic behavior that semiconductor switches are still large enough to avoid. To deal with this unreliability, biomolecules will have to be employed in "ensembles" of perhaps thousands of identical units that all simultaneously perform the same function. That way the occasional misbehaving biomolecule will be shouted down by its redundant partners, and the ensemble will on the whole function properly. The technique solves the problem neatly—but only at the cost of eliminating some of the fantastic size advantages that would be offered by a single biomolecule acting as a reliable switch.

Some say that even such biomolecular ensembles don't go far enough. To truly take advantage of biomolecules, these scientists claim, they should be employed not as substitutes for conventional computing components but rather as the foundation for entirely new computational technologies—technologies more closely based on nature's architectures.

Wayne State University computer scientist and biophysicist Michael Conrad is one researcher who thinks a biomolecular computer should operate on more nature-like principles. Biomolecules may be able to process electrical or optical signals, he says, but what proteins are really good at is recognizing and reacting to one another's shape. That's one of the keys to the brain's functioning: the neurotransmitter proteins the brain employs to transmit signals from neuron to neuron operate by fitting hand-and-glove into receptor proteins on the surface of neurons.

As a first step toward a full-scale "neuromolecular computer," Conrad proposes putting that shape-detecting ability to work in a computer-in-a-jar that could recognize patterns. In Conrad's scheme, an unidentified pattern—an image or a stream of numbers—would determine the mixture of proteins released into a reaction vessel. In an image-recognition version, for example, the solid color and long, thin shape of a pencil might release one particular set of proteins, while the curved and contrasting surfaces of a telephone would release a different set. Each of these protein sets would self-assemble into a particular "mosaic." To identify the mosaic—and thus identify the original image—the device would monitor the activity of a group of different enzymes, each one of which tends to lock onto a particular mosaic. Thus if the enzyme that targets the "pencil" mosaic becomes active, the device would know the original image was that of a pencil. With that strategy, says Conrad, "We've converted what would ordinarily be a digital signal-processing problem into a problem where we can let physics do all the work."

Another researcher intent on developing biomolecular intelligence is Stuart Hameroff, an anesthesiologist and neurological researcher at the University of Arizona. Hameroff has spent much of the past twenty years studying microtubules—long, thin cylinders of protein molecules that lend internal structure to neurons, among other cells. Microtubules "twitch," flipping an extra electron from one side of the molecule to the other, in response to various chemical and other cues; as one microtubule's twitching influences its neighbors, a signal can propagate at the speed of sound throughout a neuron's microtubule network.

Hameroff asserts that each neuron's microtubule network in effect comprises a powerful, high-speed internal computer. That may seem like processing overkill, but Hameroff and some other researchers maintain the interaction of neurons alone can't explain the capabilities of the brain. Hameroff, for example, cites the research of a scientist with Swedish car and jet manufacturer Saab who analyzed the flight of flies, concluding that there simply isn't enough processing power in the synapses of a fly's brain to account for its maneuverability. Microtubules, claims Hameroff, provide the missing intelligence. "Neurons produce a yes-or-no decision," he says, "but a lot of thinking goes into that decision. Microtubules could be the primary processing element in the brain."

If microtubules do act as a sort of computer, Hameroff thinks he knows the programming language. It takes the form of "cellular automatons," a type of information processing network in which, unlike neural networks, each node is connected only to its immediate neighbors, and these connections remain unchanging. Thus, unlike neural networks, cellular automatons can't rely on modifying connections to gain functionality. Instead, they obey a set of rules that determines how they switch from one state to another (a "state" might be a shape, a color, or a number, for example) depending on the states of its neighbors. One simple rule for a cellular automaton capable of switching between states A and B, for example, might be: switch your state if the neighbor to your right is in state A, but remain in your current state if that neighbor is in state B.

Though cellular automatons may appear to be a constrained form of computing, they have been proven in computer simulations to be capable of performing highly complex tasks. In his own simulations, Hameroff has shown that microtubular networks acting as cellular automatons—he prefers to call them "molecular automatons"—are capable of learning and adaption.

The existence of microtubular intelligence within neurons would be exceedingly bad news for any researchers who hope to construct an artificially intelligent device out of conventional computer hardware. Hameroff calculates that at the current rate of progress, semiconductor technology will produce a computer capable of human-brain-like raw processing power (though not necessarily functionality) within the next few decades—if brainpower is strictly a function of neurons acting as relatively simple switches. If each neuron has a sophisticated internal processing network based on microtubules, he claims, then the semiconductor computers of the early twenty-first century will fall short by a factor of more than a billion.

On the other hand, the discovery of microtubular intelligence would hold the promise of a new hardware vehicle for AI. "Microtubules are nature's computers," says Hameroff. "If we could understand their coding and information processing mechanisms and access them with some sort of genetic input, we could control their behavior."

To that end, Hameroff and colleagues have been modeling information flow through microtubular networks on a computer, and they are now getting ready to test parts of the model. Using a two-tipped scanning tunneling microscope, currently under development at the University of Arizona and elsewhere, they plan to electrically stimulate microtubule networks with one tip and detect how the signal propagates with the other. Once the researchers understand the effects of different signals, Hameroff intends to try "programming" microtubule networks. His long-term hope: persuading microtubules to self-assemble into structures that could serve as the basis of a self-organizing neural network. He even speaks of biomolecular "nano-robots" capable of traveling through the human body and making intricate repairs to damaged tissue. "Either by themselves, or enveloped in a cell-like membrane, assemblies of microtubules could seek out an Alzheimer's neurofibrillary tangle and destroy it with enzymes," he says. And when he really gets going, he speculates about gigantic vats of self-organizing microtubule "societies" growing in zero gravity in earth orbit, as well as about controlling people's minds by beaming in microwave energy precisely tuned to the resonant frequencies as the brain's microtubular networks.

So far, Conrad's pattern-recognition machine and Hameroff's self-assembling microtubular neural network exist only as simulations on conventional computers, though the researchers have tested some of the principles in the laboratory. "The technology has to improve a bit before we could do this sort of thing," concedes Conrad. "But if researchers can engineer proteins to act as switches, they can certainly engineer them to perform their more natural functions of shape recognition and self-assembly."

There is often an unbridgeable gap, however, between even plausible ideas and actual implementations. Is it possible that creating self-organizing biomolecular structures is beyond the scope of science? That question nagged many researchers—at least until Rasmussen's metabolism first began to take shape.

If creating synthetic life could be thought of as extreme science, then the fortyish Steen Rasmussen is the ideal perpetrator. Racing across the eerily beautiful New Mexico landscape in his Nissan 280Z, the 6-foot, 3-inch Rasmussen offers a demented Nordic grin to the blast of frigid winter air that whips through the sunroof he has opened to better enjoy the experience. "I've been reading Frankenstein lately," he shouts in his thick Danish accent above the tumult, as he reaches over to turn off the heat. "It makes you wonder about what you're getting into when you're playing around with inventing life. But someone has to do it, don't they?"

Born in Denmark and raised on a farm, Rasmussen had intended to major in biology at the Technical University of Denmark in Copenhagen when he became so absorbed in the required mathematics and physics courses that he ended up majoring in those subjects, and going on for a PhD in physics. By that time, he had already built a reputation for nonconformity. A father at nineteen, and a high-jumper who competed at the national level, Rasmussen liked to prowl up and down the coast on his thirty-foot, sixty-year-old "Dragon" racing sailboat, or hitch his infant daughter on his back and disappear into the forests on foot or ski for a day or two. Visitors to his apartment had to confront Elmer, the mounted head of a gigantic moose. "Everyone thought I was a crazy man," he recalls.

His area of specialization didn't do much to dispel the image. Inspired by the ideas of Nobel chemists Ilya Prigogine in France and Manfred Eigen in Germany, Rasmussen threw himself into the new and somewhat mysterious field of self-organization. Unfortunately, there were no established mathematical techniques for dealing with the explosion of complexity that emerges from the interactions of many simple systems. Frustrated by the lack of tools for analyzing self-organization, Rasmussen, who was by this time a physics postdoctoral student at Technical University, turned to the computer, and began creating his own simulations of abstract physical systems, an approach he called "experimental mathematics."

Soon he was often spending twelve hours a day in front of the computer, observing the way little dots following simple patterns of interaction on the screen or on printouts organized themselves into intricate patterns. It was a little like watching commodity traders at work: each individually follows a relatively simple set of instructions, but all together they can cause commodity prices to constantly zigzag wildly, as the effects of each trader's order triggers other orders, which trigger yet more orders, and so on. Seeing the same phenomenon happen on the computer with little dots was exhilarating for Rasmussen. "After being restricted to the formalism of mathematics," he explains, "going one on one with the computer and getting that immediate feedback was like taking off a straitjacket and flying."

His work baffled his fellow physics researchers, and so one of them was surprised to receive in the mail one day in 1988 a description of an upcoming conference in the United States sponsored by the Los Alamos National Laboratory and the Santa Fe Institute, a scientific think tank; the notice seemed to suggest that there were actually entire *groups* of respected scientists engaging in these weird, computer-based investigations. Recalls Rasmussen: "My friend burst into my room holding up this paper and saying, 'Look, Steen, these guys are doing stuff even crazier than you!' " Though the deadline for submitting work had passed, Rasmussen shipped off one of his papers, and quickly received an invitation to attend. Weeks later, Rasmussen had accepted a full-time joint position at the Santa Fe Institute and Los Alamos.

Incongruity seems ingrained in Santa Fe, where the mobile homes of hippie artists are plopped next to the sprawling adobe estates of wealthy industrialists. Perhaps that helps explain why at the Santa Fe Institute high-energy physicists sit rapt at talks given by biologists, and economists argue with chemists over doughnuts. The institute seems to prove its scientists' main thesis: that the whole is more than just the sum of its parts. Almost everyone at the institute studies some form of self-organization, whether it's how a protein can emerge from a soup of amino acids, or how a stock market crash can emerge from a group of stock buying and selling decisions.

Not surprisingly, Rasmussen fit right into this environment. The Santa Fe group even liked Elmer, offering to move him to the Xerox room and imbue him with speaking capabilities. (For now, at least, Elmer remains in Rasmussen's cramped Los Alamos office, sporting sunglasses). "I had felt like a nut back home," says Rasmussen. "It was such a surprise to come here and find everyone was like this."

Rasmussen immediately got to work whipping up all sorts of computer exotica meant to explore the strange world of self-organization. One of his most interesting creations is the Electronic Garden. Viewed on the computer screen, the 7100th "generation" of one such garden looks like an overhead view of a bustling city at night, complete with office buildings, heavily trafficked boulevards, and freight yards. To make his garden grow, Rasmussen has enlisted the computer's memory as "soil" and a small number of simple programming instructions as "seeds." The randomly distributed instructions, which perform such primitive tasks as pulling a small chunk of data out of one location of memory and moving it to another location, are represented on the screen as dots. Typically, the instructions end up scattering themselves uselessly around the memory. But once in a while, a group of instructions will accidentally begin to cooperate in a felicitous way, and a dynamic pattern starts to sprout; eventually, the pattern can grow into a sprawling community of cooperative programs that take over all of memory with their frenetic but highly ordered activity, resulting in the organized "city" on the screen. The addition of "noise" to the garden—that is, inserting random errors here and there—causes the flickering lights to scurry even harder in an effort to maintain their neat blocks and bustling avenues; as a result, they create even more complex forms. "Look at these little critters," says Rasmussen, beaming. "Aren't they happy?"

But while working on such abstract systems, Rasmussen was also wrestling with calculations intended to prove that the kinds of things happening on the computer were also possible in the real world. Following the ideas of Manfred Eigen, Rasmussen took the point of view that matter itself is "programmable"—that is, given a few simple rules about how matter interacts with itself, a random assortment of substances could spontaneously arrange itself into an exceedingly complex system. In other words, life itself may have sprung up in much the same way as the Electronic Garden does.

Most of Rasmussen's colleagues at Santa Fe and Los Alamos take the same view; what distinguished Rasmussen was that he became determined to prove it in a laboratory. "I couldn't count the number of people who have told me it's an absolute waste of time for a theorist to go into the lab," he says. "But I've always had my eye on bringing all this physics and math back to biology. And if you're serious about learning how biological molecules self-organize, you need to get out there and get your hands dirty."

But he wasn't quite sure how until he came into contact with the work of fellow Santa Fe scientists Doyne Farmer, Stuart Kauffman, Norman Packard, and Richard Bagley. Farmer and his colleagues were following the same sort of reasoning, but they had employed computer simulations to analyze a specific type of potentially real-life interaction: an "autocatalytic set," or a group of reactions in which each one produces a molecule that "catalyzes," or facilitates, one of the other reactions by holding two or more molecules together or ripping a molecule apart. Farmer and the others were convinced life on earth had begun with such an autocatalytic set, which would have served as a "primitive metabolism"—something that could maintain itself through an ongoing series of chemical reactions.

Farmer was proposing a process that could have emerged from simple molecules such as proteins—proteins that could have been spontaneously created in the primeval soup. Though he and his associates had only simulated the process of forming a cooperative network on the computer, he felt confident it could be achieved in the lab. But he wasn't the one who was going to do it. "We're not wet chemists," he says. "Wet," in this usage, is meant to sharply distinguish the world of real chemicals from the neater and more convenient "artificial chemistry" indulged in by Farmer. To Farmer and many other self-organization researchers, the "wet" world is simply a hopelessly messy version of the computer world. "Chemistry is another form of programming," explains Farmer, "except you have to spend a lot of time cleaning glassware."

Rasmussen didn't feel that way, and he realized that Farmer's ideas gave him the bridge to the real world he was looking for. Unfortunately, Farmer hadn't determined specifically which molecules would fill the bill, and Rasmussen is no chemist. "I hadn't done a lab experiment since high school," he says. Rasmussen is, however, infectiously enthusiastic and anything but shy, and it wasn't long before he was buttonholing researchers at Los Alamos, Santa Fe, and anywhere else he found himself to tell them his goals. One of the first to help out was Gerald Joyce, a "replication-first" researcher at the Scripps Institution of Oceanography in La Jolla, California, who suggested that Rasmussen enlist DNA for the autocatalytic set.

Of course, turning to DNA would defeat one purpose of the "metabolism-first" proposition, which is to avoid relying on implausibly complex molecules. But Rasmussen wasn't employing DNA's remarkable ability to store information and replicate itself; he just needed DNA's far simpler propensity for joining together with certain other DNA molecules. In a sense, he was tying DNA's hands behind its back to get it to stand in for a more primitive molecule. If he could demonstrate the principle of a cooperative chemical network with DNA, then it could always be attempted later with proteins. Besides, Rasmussen doesn't see a need to take sides in the metabolism versus replication debate. "My simulations show that a metabolism is easier to form than a self-replicating molecule," he says. "But perhaps they emerged together."

Soon, Rasmussen was gushing about his plan to Los Alamos biochemist Eric Fairfield. "I said, 'Steen, wait, slow down,'" recalls Fairfield. "He really needed to figure out exactly which reactions would clearly demonstrate that his ideas worked." Though already putting in sixty-hour weeks on the human genome project, Fairfield worked with Rasmussen in his spare time to plot out the step-by-step creation of the cooperative chemical network.

The first step would be to place into a beaker of salt water millions of DNA molecules—not the familiar double helix version, but the individual strands that result from pulling the helix apart. Each of these strands would comprise a chain of either adenine or thymine molecules (normally, DNA is a mixture of these two nucleotide "bases" plus two others). Then a far smaller number of longer strands of each type would be added, along with a "ligase," a chemical that binds DNA molecules together end-to-end. The vast number of short strands wouldn't do very much at first. Since adenine and thymine are "complementary" bases—meaning they have an affinity for one another—

some of the short adenine strands would snuggle up alongside short thymine strands, but then they'd quickly separate.

As for the longer strands, they would be so outnumbered by the shorter ones that they would be unlikely to find each other. However, a double-length strand might be able to attract to its side two complementary shorter strands lined up end-to-end. When that happened, the ligase would permanently fuse the two shorter strands into one double-length strand. These newly elongated strands would then float off, eventually helping to link other pairs of shorter strands together, and so on, until most of the short strands had been converted to double-length strands.

Meanwhile, the initial batch of long strands added to the beaker would have included some strands that were more than twice as long as the shortest strands. As the number of double-length strands grew, some of these extra-long strands would start to link up pairs of double-length strands into quad-length strands; as before, this reaction would start to pick up speed until the entire mixture consisted primarily of quad-length strands. And then, of course, there would be those few strands that are twice as long as the quad-length strands; these would eventually start to link two quad-length strands into an octo-length strand—and so on, and so on. Once the whole thing got going, all Rasmussen would have to do is occasionally sprinkle in some more ligase, which gradually breaks down.

Now Rasmussen had to find the time and the chemicals to realize the scheme. After all, he was moonlighting, too. "I wasn't being paid to do this," he says. "I was supposed to be building mathematical models of self-organization. I wasn't even given the money I needed to buy the chemicals." But other researchers leaked him the several hundred dollars for the chemicals by tacking it onto their own projects, and he worked after hours to make the reactions go. To measure progress, he separated out any longer strands with an electric current (DNA is electrically charged, and the longer strands lag behind) and then estimated their quantity via the radioactive tracers that had been imbedded in the short strands. All this was an exercise very different from Rasmussen's familiar computer routines, where thousands of successive generations of reactions can be simulated in a fraction of a second. "Here, I had to put something in, wait four hours, then add something else, wait overnight, add something else, wait half an hour, and so on," he says. "Sometimes I'd get busy and forget when it was time for the next step, and I'd have to start all over."

Meanwhile, he had to play around with DNA and ligase concentrations, temperature, salt content, and other variables to get the reactions to kick off. It took the Earth billions of years' worth of random "experiments" to get everything just right, but fortunately Rasmussen had an advantage: his computer simulations had told him approximately what he was looking for. Finally, the setup seemed to function smoothly. The chemical network for the first time became self-sustaining. A Proto-Metabolism was born.

Well, at least a crude first version; Rasmussen insists he's just getting started. "What I've done so far is really only to demonstrate the basic principle of an autocatalytic network," he says. "But the next phase will be much more like a primitive metabolism." That phase will include "feeding" the network a steady supply of short strands and removing the network's "waste"—that is, the longest strands. After that, Rasmussen will attempt to try to get the network to emerge from a more random mixture of DNA strands, to further bolster the plausibility of a metabolism spontaneously arising from a primeval soup. Eventually, he expects to see the network constructed from the kinds of proteins that would have been available in that soup—but he doubts he himself will be the one to do it. "Protein chemistry is just too complicated," he explains, noting that while DNA can be manipulated simply by rearranging its four building blocks, a given protein can take on radically different properties by folding into different shapes.

No matter how sophisticated the Metabolism becomes, biologists aren't likely to officially welcome it as a form of life. Notes Santa Fe researcher Chris Langton: "Biologists haven't ever had to come up with a good definition of life, because, except for viruses, it's always been obvious—something either moves around and leaves little messes, or it's a rock. They produce lists of requirements, but if we make something that meets those requirements, they'll just add something else to the list."

Rasmussen himself is equivocal. "I call it synthetic life," he says. "It's somewhere in between living and non-living." In fact, in transforming matter and energy from its environment to maintain itself, the chemical network fits the rough definition of a metabolism. Furthermore, in repeating its distinctive pattern of reactions over and over again—a very rough form of self-replication—this chemical network could be viewed as loosely meeting one of the key criteria of life offered by many biologists. (Other oft-cited criteria include adapting to the environment and evolving.) For now, that point is wide open for debate—debate that is certain to become more intense as Rasmussen and other scientists come up with increasingly complex versions of the Proto-Metabolism, and of other life-like entities.

Achieving self-organization is a major step toward creating a brain-like neural network. The next step would be to come up with a means for getting a biomolecular system not merely to self-organize, but to self-organize for intelligence. But how could self-organization be pushed in a particular direction?

Nature, of course, has a process for accomplishing just that. And nature-based AI is prepared to borrow it.

6. NATURE'S BOOTSTRAP

The theory of evolution provides a self-consistent and complete description of how life—and brains—developed. It tells us how to construct a nervous system, although the procedure is unfortunately too time-consuming to be practical.
—ANYA HURLBERT
AND TOMASO POGGIO

Chuck Taylor wants his robot, which is built out of Legos and wired up to a personal computer, to have all the behavioral sophistication of a bacterium propelling itself through a liquid in the direction of the greatest concentration of sugar. Instead of sugar, the robot's photoelectric eye detects printed stripes underneath it; the goal is to get the robot to move in the direction of the most closely packed stripes. The robot's initial program was a random collection of simple instructions that had the robot moving without rhyme or reason. But after each run, the program is slightly scrambled; if the new program is any better at the task, it replaces the original one. After several months, the robot's behavior is still sub-bacterial, but it is beginning to show signs of following patterns.

That may not seem impressive—indeed, it would take only a few hours to handwrite a program that followed the stripes perfectly—but there is good reason to believe that Taylor's approach will eventually lead to a robot bacterium that outperforms anything that could be custom-programmed. Taylor's robot is evolving.

What's more, the bacterial robot is just a warm-up. Taylor and his colleague David Jefferson at UCLA are in the midst of putting together the world's first robot farm, on which robots will be "bred" for intelligence. And even this bizarre setup will only be a modest precursor for what may be the ultimate intelligence-producing technique of nature-based AI: the directed evolution of semiartificial biomolecular "organisms."

Nature's technique for creating complex organisms, and ultimately intelligence, out of lifeless biomolecules is straightforward: get them to make copies of themselves that occasionally incorporate errors, and then wait a billion years.

What happens, of course, is that those biomolecular copies with advantageous errors—that is, errors that allow them to either replicate faster or avoid destruction longer—will tend to be more plentiful than their rivals. Sooner or later the advantageous errors take the form of slightly ever more complex entities, until eventually cell-like entities with relatively sophisticated metabolic and replicational capabilities emerge. More errors bring these cells together into the first multicellular organisms. Yet more errors produce organisms that are more and more complex, and that exhibit more and more sophisticated behaviors. In the end, you have very smart people, some of whom spend their lives trying to figure out how to accomplish what nature did, but in less time.

Can AI harness the power of evolution? In fact, evolving software has a long history. As far back as the 1950s computer scientists tried applying crude evolutionary techniques to programming. The basic approach was to take two programs each consisting of a long list of instructions, then swap chunks of instruction between them. This exchange is analogous to the "crossover" that occurs in nature when an egg and sperm fuse: their chromosomes line up and swap genetic material so that the resulting fetus will have genes from each parent. Even cellular organisms that reproduce through cloning exchange genetic material. The exchange serves evolution in two ways: it allows the occasional creation of offspring that exhibit the best qualities of both parents, and (more rarely) it provides an opportunity for advantageous errors, or "mutations," to creep in.

Unfortunately, what works beautifully in nature was a disaster for computers. The problem was that computer programs must be syntactically precise: that is, there is little leeway in the manner in which instructions can be grouped, and in the order in which they appear within a group. If any scrambling occurs, the program is rendered thoroughly meaningless, and the computer refuses to run it. Researchers tried to get around the problem by restricting the use of instructions in such a way as to increase the chance that usable mixings would occur, but the resulting hobbled programs were of little interest.

The first breakthrough came from John Holland, a University of Michigan psychologist and computer scientist, who after a decade of effort came up with a technique in the mid-1970s for representing programs as strings of 1's and 0's. These DNA-like "bitstrings" essentially coded information as a series of yes-or-no questions: for example, the third bit in a string might represent the question, "Should these two numbers be added?", with a 0 representing no and a 1 representing yes. These "classifier systems," as Holland called them, readily lent themselves to the process of crossover and mutation, since the program could be designed to work for all combinations of 1's and 0's. What's more, Holland proved that virtually any program could, at least in theory, be represented as a classifier system.

Holland's classifier systems evolved nicely. A small initial "population" of strings consisting of random combinations of 1's and 0's might all be useless in solving a particular problem. But after "mating" various pairs of the population and carrying out the crossover (and throwing in a few random mutations), a few strings might show a little promise. These superior strings would then serve as the parents for the next "generation," and this process would be repeated for anywhere from dozens to thousands of generations. In the end, Holland's "evolving algorithms" usually produced a bitstring that did the job, without any purposeful programming having been undertaken.

Holland's technique was particularly useful for problems with many variables, each of which has many possible values, so that manually examining each possible solution would be prohibitively time-consuming. In such cases, genetic algorithms can be unleashed to "explore" the possible solutions. The technique has been successfully applied to a number of real-world problems, including solving mathematical equations, designing a control scheme for a gas pipeline system, and designing communications networks. General Electric has even come up with jet engine turbine designs through the use of genetic algorithms.

The big drawback to Holland's technique is that converting a conventional program into bitstring form is usually a prohibitively time-consuming process. Coming up with a way to evolve programs without first converting them was the task undertaken by John Koza in the 1980s. Koza studied under Holland in the 1960s, and then left academia to cofound Scientific Games, a company that produced state and national lottery systems. When Koza left the company in 1987, having sold it a few years earlier to Bally, annual sales were $86 million, and the company's systems had produced 20 billion lottery tickets for virtually every state in the United States and a number of other countries.

The now outrageously wealthy Koza immediately founded a high-tech venture capital company, and settled into a spectacular home overlooking Silicon Valley. (The driveway is so long and winding, and the dropoff at the end so abrupt, that Koza had to install a roadsign near the top warning of the cliff ahead.) The starkly contemporary decor of Koza's home, broken up only by the many Leroy Neiman paintings, yields no hint of Koza's high-tech background. But tucked away in a separate section of the house is a large room that looks like the data center of a midsized bank. It is there that Koza indulges in what is essentially a hobby for him: evolving software.

Koza thought Holland had the right idea, but felt the technique would never reach its potential unless it could be applied to whole programs. But how could the syntax problem be overcome? "If you can genetically breed a computer program to solve problems, then you can solve all problems," says Koza. "But people think of computer programs as brittle and unforgiving; you put a minus sign instead of a plus sign, or a .4 instead of 4, and the program doesn't work."

The answer, Koza decided, was to modify a programming language in such as a way as to allow programs to be chopped up without killing them. The language he picked was LISP, which had been invented by McCarthy in the 1960s specifically for artificial intelligence. LISP, which stands for List Processing, allows the flow of a program to go branching off in many directions, instead of requiring, as most languages do, that the program logic be followed in a serial, step-by-step fashion. Koza took advantage of this branching to create a genetic crossover technique that always swapped chunks of programs at corresponding "branch points." Since the chunks of programs between these points could be made fairly self-contained, swapping them in whole pieces wouldn't destroy the program—and once in a while, it would improve it.

Otherwise, Koza's technique works much the same way as Holland's: an initial population of randomly constructed programs—Koza typically begins with more than ten thousand—has a go at a problem, the ones that do the best are mated to provide a new population, and the process is repeated until a program that solves the problem is produced. Interestingly, the programs that result from Koza's technique look jumbled and illogical to a human programmer's eye, even though they typically work better than a handwritten program. That's because evolution works in a different style than humans do. "When a human programs," he explains, "the initial versions usually don't work at all, and then suddenly he'll come up with a version that works perfectly. But nature goes through a succession of steps, improving gradually. So the trajectory of genetic programs consists of weird expressions that get closer to the value of what you want, without looking anything like human programming."

Koza's technique doesn't necessarily perform better than Holland's, but it can be applied to a much broader range of problems. Koza recently published a book of seventy "benchmark" problems that are well known to the computing science community—one problem, for example, involves producing a program that will back a truck up—and shows how his genetic algorithm technique produces either better solutions or produces standard solutions with less human effort.

But the fact that genetic algorithms work well is only one-half of the appeal of this technique to nature-based AI. The other half lies with the fact that it happens to be nature's technique, which makes it a preferred path for arriving at intelligence. Not surprisingly, then, more and more AI researchers are trying their hands at artificial evolution.

• • •

Pattie Maes, the Belgian scientist who as a member of Rodney Brooks's group programmed Attila to teach itself to walk, left the MIT AI Lab in 1991 to join the school's Media Lab, which explores, among other things, ways to improve the interaction between computers and people. Though she still does some work with Brooks's mobile robots, Maes spends most of her time on what she calls "intelligent agents"—programs that simulate some of the behaviors and even appearances of living creatures, but that are designed to serve a useful role for their computer-user "owners." Maes claims there are two possible approaches to enabling AI to have a bigger impact on people's lives: "You can bring computers into people's environments with robotics," she says, "or you can bring people into the computer's environment. I believe the second approach is more feasible, at least in the short term. People are going to spend more and more time in virtual environments."

The idea of the computer user entering a "virtual reality," or "cyberspace," in which the user temporarily feels more grounded in the world created by the computer than in the real world, is typically associated with electronic goggles, gloves, and even full suits that contribute to the illusion of "moving" through computer-generated images. But Maes envisions a less literal but perhaps equally compelling cyberspace that achieves its effect not by sensually encompassing the user but rather by getting users to enter into the same sorts of (nonphysical) relationships with intelligent agents as they do with people and animals.

One of Maes's intelligent agents, for example, looks and acts like a dog. When the dog-agent first appears on a new user's screen, it might act afraid of the user, running from the mouse-controlled pointer until the user "feeds" the dog by dragging a picture of food to it across the screen. After a few feedings, the dog starts to become affectionate and crave attention, though it may growl at new users. The dog can learn to do chores, such as "fetching" icons (pictures representing data or computer functions) needed by the user. It can also help the user interact with users on other computers by, for example, grabbing an electronic mail message in its mouth, running with it off the screen, and appearing (having "traveled" across a computer network) on another user's screen to drop off the message. Dogs could even run to another computer's "backyard" to play with another dog, returning home when summoned by the user, or coming back on its own to take a nap.

Another Maes agent is more human in form and behavior, resembling a sort of squared-off happy face. In one application, the agent serves as a calendar-arranging assistant. After spending time observing the way in which the user sets up appointments to learn his or her priorities, the agent attempts to take over the chore, intercepting e-mail messages requesting meetings and placing them on the calendar. The face keeps the user apprised of its efforts through a wide variety of facial expressions: it screws its face up, for example, while it mulls over a difficult decision, grins if the user approves of its decision, and looks shocked if the user overrides its decision (though it will learn from the rejection).

To make her agents smarter and more customized to individual users' preferences, Maes has turned to artificial evolution. To "breed" a better electronic mail sorting agent, for example, Maes starts with a small "population" of mail "retrievers," represented on-screen as golden retrievers—partly as a visual pun but also, claims Maes, to help users relate the bizarre concept of evolving software to the more familiar one of dog breeding. Each retriever uses a different set of priorities to sort incoming electronic mail: one might tag all articles with sports terms as "must-read" items, while another would (more reasonably) give preference to e-mail letters sent from the user's boss. After examining the retrievers' work, the user indicates which ones did the best job. Those few are then mated: their offspring are provided with some characteristics from each parent, along with a few random mutations. The offspring replace the rejected retrievers, and the process is repeated until the user is satisfied with a particular retriever's work. Maes claims the users who have tried the retrievers all report having been able to cut down on time spent reading "junk" e-mail without missing important messages.

Such intelligent agents, insists Maes, will be seen by users not only as useful and entertaining, enlivening otherwise dreary computer chores, but also as endearing, and even lovable. "As computers become more important as a medium for social interaction, intelligent agents will be equally or more important as robots with real bodies," she says. "Agents in a computer could even have as much effect as agents in the real world." Maes has discussed her work with various companies in the entertainment industry, some of which apparently see a bright future for interactive video systems containing entire populations of life-like artificial animals and people, as well as with mainstream software companies seeking to make their programs "friendlier."

Maes is attempting to approach some features of higher-level intelligence in her work. However, most researchers interested in achieving AI through evolution are doing so according to the first principle of nature-based AI: start with the simplest behaviors and work up. But in attempting to evolve simple behaviors, AI has some competition in the form of a brand-new science: artificial life.

Artificial life researchers create all manner of animals, plants, cells, abstract life forms, and often entire communities or "ecologies" boasting a wide range of interacting, evolving "life-forms"—all on a computer. These electronic creatures are typically represented as video-game-like forms on a screen that mill about, chase each other around, hunt for "food" (other blobs on a screen), mate, and otherwise perform the basic functions of life. Unlike a video game, however, these behaviors are not specifically programmed in; instead, scientists simply provide them with rules of interaction, and then sit back and watch what happens. Thus, for example, one computer creature might be told to move toward food, and move away from other creatures—it's anyone's guess as to what direction this creature will move in when it's surrounded by both food and other creatures.

Evolution plays a large role in most "A-life" projects. As the simple creatures play out their roles, those that are the most fit—that is, those that, for example, find the most "food," best survive predators, or display characteristics that are of most appeal to potential mates—mix and match their "chromosomes" (typically Holland-style classifier system bitstrings that encode their characteristics) and undergo mutations, creating offspring to take the place of less successful creatures.

Because artificial life's denizens can be made to live out their lives and reproduce in millionths of a second on high-speed computers, scientists can in a few hours observe patterns that require millions of years to unfold in real life. The hope of most artificial life researchers is that not only will they come up with observations that closely mimic nature's known processes but that they might be able to make observations that enable predicting other natural processes that have never before been identified. There has been encouraging progress. In the late 1980s, for example, Kristian Lindgren, a researcher at the Chalmers University of Technology in Sweden, noticed that the evolving abstract life-forms in one of his programs seemed to go through long periods of little change, separated by puzzlingly brief periods of intense evolutionary change. The program, he suddenly realized, was providing a precise imitation of a now-favored model of natural evolution known as "punctuated equilibrium." Lindgren hadn't programmed the behavior in; it had spontaneously emerged, just as, presumably, it had spontaneously emerged in nature.

One of the most visible researchers in the A-life community is the Santa Fe Institute's Chris Langton, a friendly but intense and rough-hewn figure in jeans and camping vest who even at an organization where a bolo tie passes for formal attire looks like someone who has walked in off a construction site. Langton is one of the most passionate advocates of the "strong claim" for artificial life's validity. The weak claim is controversial enough: it contends that in doing good simulations of life, you can learn about real life. Many people find that assertion hard to swallow, insisting that real life depends on far more variables than a computer program could take into account. No wonder, then, that these doubters positively choke on the strong claim, which, simply put, insists that artificial life *is* real life. "Life is a process, and it shouldn't matter what the hardware is," says Langton. "If a simulation meets the criteria for life put forth by biologists—that it can maintain itself, that it self-replicates, that it evolves—then it should be valid to ask if it's alive."

While most scientists, and even many A-lifers, regard Langton's strong claim with skepticism, the leap to the weak claim is not nearly as large, especially for those who have been working with genetic algorithms to begin with. John Holland himself, for example, has in recent years turned his attention to creating simulated worlds in which evolving A-life denizens come to display such complex phenomena as group cooperation, "arms wars," mimicry (in which creatures scare off predators by fooling them into thinking they are better-defended), and the development of multiple, interdependent species.

It's not just computer scientists who are enthusiastic about A-life. Thomas Ray is one of a number of biologists who think A-life provides an invaluable laboratory for exploring the subtleties of biological evolution and biodiversity. Ray first thought of the idea of artificial evolution in 1978 when, as a student at Harvard, he was playing a game of Go, a Japanese board game in which the pebble-like pieces placed by the two players form constantly changing, interactive patterns in a competition to proliferate. The thought stayed in the back of his mind for ten years, until he acquired his first personal computer. As he started to experiment with writing programs for it, he realized the machine would make a perfect "virtual jungle" on which to try out his ideas.

Ray had soon written a simple program that did little but replicate itself, occasionally making an error that served as a mutation. When he unleashed the program into the computer's memory, it quickly proliferated, and then evolved into separate populations of "hosts," which gathered information from the "world" around them, and "parasites," which stole information from the hosts. The parasites thrived to the point where they nearly wiped out the hosts; but with a lack of hosts, the parasites started dying off, allowing the hosts to return, which renewed the parasite population, and so on, back and forth. Soon hosts and parasites were developing an ongoing series of alternating strategies until a sort of balance was reached. Ray's creatures also discovered sex: they were exchanging their genetic codes, allowing them to evolve independently of mutations. All of this happened the first night Ray ran his program.

Even the military is interested in A-Life. John Grefenstette at the Naval Research Laboratory, for example, is evolving simulations of jet fighters that are becoming more and more adept at dogfights. Theoretically, the more highly evolved versions could be gleaned for new combat techniques—or even adapted to help pilot a real fighter jet.

There are obvious similarities between A-life and AI. Both are new sciences outside the mainstream, both run on computers, both are biologically inspired, and both aim to achieve behaviors considered entirely unique to living beings. Even A-life's "strong" and "weak" claims have hotly debated analogies in AI: can a machine become truly intelligent, in the sense that it would become conscious, or can it only simulate intelligence?

But there are also differences between the fields. Perhaps the most significant one is that AI researchers have a goal in mind for their programs: they want them to produce behaviors considered intelligent. On the other hand, in A-life, as in nature, evolution is aimless; it simply produces a wide range of variations, and those variations that survive longer or reproduce faster tend to stick around, while other variations disappear. Thus A-life concerns itself with whatever thrives among the variety, while AI is interested only in those varieties that are smart.

But the difference may be more superficial than it at first seems. With intelligence having come so late to the evolutionary party and playing such a large role in our own existence, we tend to think of it as representing a "higher" stage of some sort of evolutionary progression. But as University of Michigan AI researcher Stephen Kaplan has said, "The only reason we're so smart is that all the good niches were taken." In other words, intelligence perhaps isn't so much a refinement of evolution's earlier efforts as it is a consolation prize, a way for a naked ape to survive in a world filled with hardier and faster-proliferating creatures such as bacteria and insects.

Most researchers agree there is plenty of reason to believe the two fields should overlap. Intelligence, after all, is a product of evolution; and any behavior that enhances the prospects of survival or reproduction could be considered a form of intelligence, if a low-level one. In a sense, biological life and intelligence depend on one another, and A-life and AI may prove to be equally codependent. Stephen Grossberg, for one, acknowledges that intelligence removed from the context of the development of life is a sterile concept: "One of the main things governing the evolution of the brain is behavioral success," he says. "Who gives a shit if you have beautiful neurons if they can't adapt to the environment? There would be no benefit, because you'd be dead."

Likewise, Langton sees the AI/A-life line as a blurry one. "Our goal is not explicitly to get to intelligence," he says. "But there may be something about the low-level building blocks of life that suggests new ways to come at the problems of intelligence, to work up from there toward reasoning and planning instead of the other way around." Langton's sentiments are, of course, precisely those of Rodney Brooks, Stewart Wilson, and the entire "animat" movement, which attempts to achieve low-level intelligence in robots governed by relatively simple, biologically inspired behaviors. Thus the line between the two fields has blurred, particularly in the realm of artificial evolution.

The animat movement has made significant progress since Brooks introduced his subsumption architecture in 1986, and even since the development of Attila in 1991. Hundreds of researchers now dedicate themselves to studying "ethology"—animal behavior—and incorporating whatever principles they can glean into the design of control systems for real or simulated robots. The result is robots that can perform a wide variety of surprisingly complex functions.

Michael Arbib and Hyun Bong Lee at the University of Southern California, for example, have enhanced their robot's obstacle avoidance abilities by studying the way in which frogs attempt to snag a fly on the other side of a picket fence. Thomas Ulrich Vogel at the University of Cambridge has developed a two-legged robot that teaches itself to step over fences and other obstacles—it "walks like a person on stilts, a little bit like Charlie Chaplin," claims Vogel. Robb Lovell and others at Arizona State University have designed an animat-style control system for an underwater robot capable of making its way through currents and eddies in pursuit of food.

In technical violation of Brooksian orthodoxy, a number of researchers have attempted to imbue robots with simple mechanisms for tracking their positions by constructing rudimentary internal maps of landmarks and then gauging their progress against these landmarks. And one of the most active areas of animat research involves the challenge of building multirobot "societies," or "swarm intelligence," in which robots cooperate on tasks such as cleanups and modest construction chores. Most of these robot societies, like the five-robot group constructed by the University of Alberta's Ronald Kube and Hong Zhang after studying insect societies, coordinate their tasks not through language but rather by having one robot's behavior trigger complementary behaviors in another.

One of the first animat researchers to turn to A-life-style artificial evolution was Stewart Wilson. Wilson has remained a central figure in the animat movement, having coorganized two international conferences in the rapidly growing field. Much of his work has centered on determining ways to standardize the simulated environments in which simulated animats function, so as to provide researchers with an objective means of determining the relative strengths and weaknesses of their creations. But Wilson has also applied John Holland's genetic algorithm techniques to animat behaviors, and even to the simulated "biological" development of new animat species. "Creatures ought to be able to gradually improve their abilities beyond what they can learn in their lifetime," he says. "Evolution is a method tested by nature for a very long time."

The animat evolution schemes employed by researchers are straightforward, though the goals vary. Marco Dorigo of the International Computer Science Institute in Berkeley, for example, has evolved a control program that enables his "AutonoMouse" robot to effectively chase "prey" or escape a "predator." Tetsuya Higuchi in Japan's Electrotechnical Laboratory at Tsukuba evolved a foraging behavior that mimics, among other things, the tendency of real animals to broaden their menu when preferred foods become scarcer. And Craig Reynolds with software vendor Electronic Arts developed a group of simulated "critters" that evolved a herding instinct to avoid predators.

John Koza has also dabbled in robot control programs, in some cases evolving new versions of programs already developed by other means, just to prove the effectiveness of his genetic algorithm technique. He took one robot wall-following program developed by hand in twenty iterations over a period of months by MIT's Maja Mataric, for example, and reproduced a slightly better-performing version of it in a matter of hours by starting with a population of 1,000 random versions and letting them evolve for fifty generations. Koza firmly believes that most AI researchers will turn to evolution sooner or later; even Brooks and his cohorts, he predicts, will become converts. "There's a 90 percent chance they'll abandon reinforcement learning and end up at genetics within two years," he says. "It's the logical way to go. After all, we know more about chromosomes than we do about how the brain is structured."

Of course, neural network researchers are learning more and more about building brain-like structures, which suggests that a particularly effective path to intelligent animats might be to apply evolution to neural-network-based control programs. Nature clearly works that way, having relied on evolution not merely to develop the mechanics of brains but also to figure out much of an animal's initial wiring. A fetus's brain starts to make connections and set weights according to a scheme honed by evolution. It is only when we become drenched in sensory inputs after birth that the wiring process starts to shift from nature to nurture—but by that time our evolutionary "hard-wiring" has already had a large influence over what our brains will be prepared to learn. In fact, neural wiring may be one of evolution's most dynamic activities. Recent studies have shown that the brain of the domestic cat exhibits substantial differences from the Spanish wildcat—the former has had as many as two-thirds of some types of neurons "pruned" away—even though the two are separated by a mere 20,000 years of evolution, a time period previously thought to be too short to allow for significant evolutionary change.

Some animat researchers have been inspired by such biological role models to experiment with evolving neural networks. Neural networks themselves are nothing new to the animat movement. Of the many researchers working in this vein, perhaps the best-known is Randy Beer at Case Western Reserve University. Beer developed a control system for a cockroach-like robot that resembles Brooks's efforts in its reflexive, behavior-based approach, but instead of interacting miniprograms Beer's cockroach is run by a neural network. More recently, Beer has employed genetic algorithms to, in effect, evolve the optimal weight settings for the connections between nodes, resulting in increasingly sophisticated behaviors—especially in the coordination of the six legs—that would have been otherwise difficult to achieve. Other researchers have used evolving neural networks to enable robots to navigate around obstacle-littered fields, and to develop cooperative behaviors with other robots. Wilson has even evolved the connections to multilayered perceptrons capable of solving a range of difficult problems (and demonstrating yet again that Minsky and the rest of the AI community was premature in writing perceptrons off).

UCLA artificial intelligence researcher Michael Dyer has also been active in evolving neural networks. Dyer, a self-proclaimed "ex-symbol-pusher" who studied under neural-net detractor Roger Schank, of all people, has more than made up for lost time by looking at neural networks from a dizzying variety of perspectives. One of his neural networks, for example, is designed to indulge in a computer version of daydreaming. "Computers never think about how things might have been," he explains. "But humans are always mulling over alternative pasts and futures, like how you'd tell your boss to go to hell if he told you you couldn't have a raise." Another Dyer neural net tackles morality; he says he would like it to understand, for example, that just because hitting an old lady over the head might be the most effective means for getting money in a hurry doesn't mean it is the best means for getting money.

If Dyer can be said to have a particular specialty within neural networks, it is how neural nets might be capable of developing language capabilities, and how such capabilities are related to thought. It's not an easy subject to tackle in the context of artificial neural nets, since neuroscientists haven't yet figured out how people develop and apply language skills. "Where does the human brain encode the concept of a word like 'irresponsibility'?" he asks. "We don't even know where we keep a word like 'chair.'" Developing artificially intelligent systems with such advanced capabilities, he claims, requires an ascending hierarchy of AI techniques. Symbolic approaches to AI occupy the lowest rung of Dyer's hierarchy; each step up brings AI toward systems that have more and more nodes with greater and greater interconnectivity. At the top of his hierarchy are evolving neural networks.

Dyer has taken an animat-style approach to evolving language capabilities on neural networks. He starts with a population of neural-network-based creatures, and places them in an extraordinarily rich virtual world he calls "Bio-Land." In Bio-Land, his "biots" can use their senses of sight, sound, and even smell and touch to find food and locate mates. Biots hunt in packs, bring food back to their children, and huddle together for warmth. What Dyer really wants his biots to do, though, is talk among themselves; he hopes they will evolve a primitive language. "They're not going to sit around and evolve the language skills to argue about politics," he says. "But if one of them developed a signal that allowed it to alert others to food when it found some, it might give it a selective advantage that would carry into future generations." From such a primitive language, speculates Dyer, higher levels of thought might eventually arise.

Thinking Machines founder Danny Hillis is also a booster of the evolutionary approach. He spends much of his time developing software that evolves into host and parasite species, as did Thomas Ray's program. Hillis specifically wants the parasites, though; he believes—as do many evolutionary biologists—that the need to develop resistance to parasites is one of the driving forces behind evolution. Urged on by ever-evolving parasites, Hillis's software may, he hopes, solve a range of practical problems, and ultimately become artificially intelligent. Apparently, hardware is no longer a barrier: Hillis asserts his latest Connection Machine, the CM-5, has just about enough nodes and raw processing power to evolve into a computer version of a human brain. (Hillis has famously said that his goal is to build a computer "that will be proud of him.")

Animat researchers are working on a number of ways to improve the effectiveness of evolving software for creature-like robots. One obvious approach is to increase the initial population size of the evolving programs, which would increase the chances that some programs will luck into mutations or genetic swaps that prove advantageous. Unfortunately, running more and more creatures simultaneously on a computer becomes, as Wilson puts it, "multiplicatively complex"; limited by time and processing power, researchers are often forced to make do with perhaps a few hundred simulated creatures, compared to the countless trillions available to nature. Running the programs for an extended number of generations only partially compensates for the small populations, so researchers are looking into ways either to simulate larger populations more efficiently or to harness more processing power.

Another advantage nature has over artificial evolution is "biological development," the process by which the genetic code embodied in a set of DNA molecules in a fertilized egg unfurls in a multistage growth process into a fully formed organism. Buckminster Fuller used to refer to the "trim tab" effect: a small force applied to a wheel at the helm of an ocean liner moves a tiny rudder—a trim tab—whose action moves a larger rudder, whose action moves the giant rudder that turns the ship. In the same way, biological development acts as a sort of trim tab, or complexity amplifier, for evolutionary change: a single, simple change in a DNA molecule of a fertilized egg can set off a chain reaction of change as the egg divides and the organism takes shape, perhaps ultimately resulting in a more efficient brain, or stronger wings. In animat evolution, on the other hand, evolutionary change is more direct: to effect a given improvement in an animat, which is essentially "born" fully grown, the software must explicitly make that improvement. Thus animat evolution tends to proceed in smaller steps, with fewer opportunities for the sort of wild improvisation that nature achieves. To add some of this complexity to artificial evolution, Wilson has been working on animats that grow at "conception" from single- to many-"celled" organisms, but the efforts are in an early stage.

Richard Belew, a computer scientist at the University of California at San Diego, has suggested adding more group dynamics to animat simulations, so as to inject "tribal"-level evolution into the populations. Evolutionary changes often propagate within a group of organisms, he notes, and not across the species as a whole; it is then competition among these groups, or tribes, that determines which changes will ultimately be promulgated through the species. Belew has also explored the idea of adding the capability to develop "cultures" to animat populations, as a sort of bridge between the learning approach and the evolutionary approach to AI. "Culture accumulates the 'wisdom' of individuals' learning beyond the lifetime of any one individual but adapts more responsively than the pace of evolution allows," he asserts.

But even if all these possible improvements were to be implemented, animat evolution would still fall short in one critical aspect: it takes place in simulation on a computer, and not in the real world. Because it is anchored in a simulated world, animat software evolves to produce behaviors that allow it to deal superbly with simulated obstacles, simulated landmarks, simulated sounds, and simulated tasks. But when transferred to a real robot in the real world—ostensibly the goal of most animat researchers—the software usually falls flat, in some cases quite literally. To obtain effective real-world behaviors, animats should ideally be evolved as real robots. "Work with simulations doesn't carry over to real machines," says Pattie Maes, who has worked extensively in both arenas. "Using robots forces you to find methods that really work in an environment where things break down and sensors are unreliable."

But the idea of developing evolutionary software directly on robots is a daunting one. For one thing, robots are harder to operate than simulations, which is why most animat researchers depend on simulations in the first place. "One of the reasons I'm very happy to not be working with robots anymore is because it's so frustrating," says Maes. "You end up spending 90 percent of your time on the machinery and electronics, and in general on what I consider to be non-interesting problems." For another thing, how does one work out the mechanics of breeding robots? And if working with a single robot is frustrating, what would it be like to deal with a population of robots? "Nature has fruit flies to work this stuff out on," says John Koza. "Nobody's got a thousand robots."

No, but David Jefferson and Chuck Taylor are planning to get a hundred.

"Robot software will be by far the most difficult and complex software anybody has ever built," says Jefferson. "If you look at all the elements that can make software hard to develop, this has them all: they operate in real time, their motions are complicated, their electromechanical systems differ from robot to robot, you can't predict the different conditions each will face, and a single software system has to work with all of them. Not only that, but bad robot software can result in harm to both the robot and its environment."

The closest thing to robot software in complexity is the software that controls spacecraft, and that has required tens of thousands of programmers writing and debugging programs for decades—and spacecraft usually operate in groups of one. Jefferson, meanwhile, wants to build robots that learn to work cooperatively en masse, marching together, lifting heavy objects, even constructing things. "If you look at all the examples we have of complex cooperative behavior in the world," he says, "you'll note that none of them have been engineered; they're all in the animal kingdom, which means they've all evolved."

Jefferson is a computer scientist who, along with graduate student Rob Collins, specializes in developing computer simulations of evolution. The two have also created simulations of ant populations, in which colonies that at first wander around randomly searching for bits of food evolve the ability to organize their searches by laying scent trails—just as real ants do.

Along the way, these simulations sometimes reveal patterns of evolution that suggest answers to riddles that have been puzzling evolutionary biologists for decades. For example, why does nature favor sexual reproduction over cloning? Some biologists have suspected that the gene-scrambling process of conception makes it easier for a species to stay one step ahead of viruses, which tend to target themselves to a particular gene structure. And that's exactly what Jefferson and Collins have seen on simulations in which digital viruses go after populations of maters and cloners.

Taylor comes at AI from the other side: he's a biologist who turned to evolving computer simulations to predict how populations of various organisms evolve and grow. Take the mosquito, which carries diseases that rob the lives of 200 children worldwide every hour. The best way to control mosquitoes would be to spray prime breeding grounds, but biologists have never been able to predict exactly where these breeding grounds will spring up—until Taylor started employing techniques similar to Jefferson's to evolve computer simulations of the growth of mosquito populations. The breeding grounds of his digital insects now match up to real-life breeding grounds with as much as 95 percent accuracy, and he has been working with health authorities in California and Africa to put his simulations to work in mosquito control efforts.

Buoyed by their respective successes, and looking for a new challenge to which they could apply artificial evolution, Jefferson and Taylor started to think about robots as a way of bridging the gap between the idealized, flat world of computer simulations and the hopelessly complicated world of real-life organisms. In fact, the planned robot evolution process is basically the same as the process employed for evolving the ant simulations; the key difference is the ants roam in a relatively predictable virtual world, while robots perform in the real one. Of course, that's a significant issue, as witness the slow progress being made by Taylor's bacterial robot.

But Taylor points out that evolution is a bootstrapping process: it's difficult to kick it off, but once it's rolling it works great. Getting the process kicked off is a problem Jefferson has run into with his and Collins's ant simulations. "The hardest part is at the beginning of evolution," he explains. "When you start with totally random behavior, only a small fraction of ant colonies ever accidentally starts to gather any food at all. But once you get one generation that starts, they're likely to go on to evolve an effective strategy." In addition, evolutionary processes don't work well with a population of one; desirable traits are much more likely to emerge in reasonable time if there is a large population among which to mix and match characteristics.

That's why Jefferson and Taylor are setting up a robot farm. They're planning to start with a herd of 20 robots, and then move within a year to 100. "Actually, I'd like to have tens of thousands, or hundreds of thousands of robots," says Jefferson. "But nobody could afford that."

To breed robots, Jefferson and Taylor will equip each member of the herd with a randomly scrambled program for performing a task, such as gathering Ping-Pong balls. The poor-performing robots would all be "killed," meaning their programs will be wiped out (as of yet, Jefferson and Taylor do not have robots' rights activists to worry about). But these machines will not have died in vain. A few lucky robots would have ended up with a combination of instructions that might have brought them at least partway along the task at hand. Two of these better-performing robots would then be designated "parents," and they would "mate"; that is, a crossover would be performed on their programs, with an occasional mutation thrown in. The resulting new programs would then be inserted into the microprocessor brains of the many robots that had been killed off—a form of robot reincarnation. In this way an entire new generation of robot programs based on programs of the talented parents will be created and let loose. Most of these second-generation versions will probably end up doing worse than the parents. But a few will probably do better, leading to a new set of parents for the third generation. And so on.

In practice, Jefferson and Taylor plan to run through as many as thirty or so generations every eight hours, watching each generation just long enough to determine the best candidates for parents. With each successive generation, the robots should get more and more adept at their assigned task, as natural selection—or, rather, artificial natural selection—takes its course. "Eventually we hope to get to a state where we can consider those robots to be a colony of pseudo organisms," says Jefferson.

One concern is that the planned population of 100 robots won't climb the evolutionary tree quickly enough to make things interesting. To get things rolling, Jefferson may pad the robot population with computer simulations of several thousand additional robots. The simulations will speed the process up, but the real robots will keep things honest. "There are a huge number of ways in which a simulation can't accurately reflect the situation of real robots in a real, 'dirty' environment with dust, and cracks in the floor, and mechanical failures," he explains. "After all, our own genetic software in nature was created in the context of real organisms that have to survive in the real world, so the scientific point is blunted in simulation."

The most immediate obstacle to the robot farm is money. Robots are expensive—about $7,000 each for MIT-designed, toy-tractor-like machines that Jefferson and Taylor have their eyes on—but the two are optimistic the funding they've applied for will come through. But even then, Jefferson will first face the task of writing a programming language, control programs, and other basic software the robots will need even before they start evolving. "A huge amount of software infrastructure has to be built before the robots are convenient to use," he says.

If the evolving robot project works, the techniques could be applied in other areas. For example, Jefferson envisions automated roadways on which robot vehicles cross paths at high speeds without risk of collision, all under the control of evolved software. But the beauty of the evolutionary approach—that it emerges on its own, without anyone's having to analyze all the complexities—is also a potential stumbling block to its acceptance. "The software may function perfectly well, but we won't really understand how it works," explains Jefferson. "Especially if human life is at stake, legal liability probably wouldn't allow evolutionary programs without a stronger guarantee that they can't fail."

Can artificial evolution be applied to biomolecules, and perhaps eventually to a biomolecular intelligence? If so, most researchers believe that RNA is the biomolecule to begin with: a popular theory that holds that a self-replicating strand of RNA was a precursor to the more complex, DNA-based forms of life familiar to biologists. "It's generally believed that there was a time roughly four billion years ago when RNA was running the show," says Scripps Gerald Joyce, a leading RNA researcher. "We just don't know how an RNA life-form came about."

Indeed, just how nature managed to come up with the first prodigious strand of self-replicating RNA has been a deep and longstanding mystery. Even if one allows that some form of RNA somehow sprang up whole out of the primeval soup, it shouldn't have been able to do anything useful. That's because in nature, RNA (like DNA) relies on "enzymes" for replication. These complex molecules enfold the strand of RNA and, using that original strand as a "template," string together building-block molecules called "nucleotides" into a "complementary" strand, which is a sort of photographic negative of the original strand. The enzymes then repeat the process on the complementary strand to construct a precise copy of the original.

The enzymes, meanwhile, are made up of proteins, which are in turn assembled according to molecular instructions embedded in RNA. In other words, RNA directs the assembly of proteins that form the enzymes that enable RNA to be duplicated. It's a nice system—but how did it get started? Without enzymes, it would seem that RNA couldn't have copied itself and evolved; but without highly evolved RNA, you don't have any enzymes. It is biology's ultimate chicken-and-egg question.

In the 1960s, researchers posited an elegant way out: a primitive RNA molecule that could also act as an enzyme in its own replication. Two of these identical molecules acting in concert—one as an enzyme, the other as a template—might have been able to churn out a third copy unaided. The molecule then could have eventually evolved into a more complex version capable of synthesizing enzymes to carry out the job more efficiently. And the rest, as Darwin might have said, is natural history.

The only problem was that no one had ever found a form of RNA that could do double-duty as an enzyme, relegating this hypothetical molecule to the status of evolutionary biochemistry's missing link, and leaving that particular origin-of-life scenario no more than an interesting speculation. But then, in 1982, biochemist Tom Cech dug into the cells of a single-celled, paramecium-like creature found in ponds called *Tetrahymena* and came out with a strand of RNA capable of performing some simple enzymatic functions. That discovery earned Cech a Nobel Prize in 1989, and seemed to give proponents of the self-replicating-RNA theory the boost they had been looking for.

But the sense of triumph was short-lived. "When RNA enzymes were first discovered," says Joyce, "there was a flurry of papers and commentary saying, 'Well, that solves it, we can see there was an RNA life-form.' But the difference between what these molecules could do and self-replication is actually quite significant." Specifically, it had quickly become clear that Cech's RNA, along with the eighty or so other versions of enzymatic RNA from a variety of microorganisms, plants, and fungi discovered in the next few years, couldn't manage much more than hacking apart their own strands in a few specific places, and perhaps sticking on a few nucleotides at the strand's end. It was an enzyme's handiwork, to be sure, but it was a far cry from replication.

Jack Szostak was in his early thirties, studying and manipulating the DNA in yeast cells, when word of Cech's discovery reached him. Suddenly, Szostak knew what he wanted to do with the rest of his life. He was going to hunt for an RNA enzyme that could be coaxed into self-replication.

Szostak hadn't picked the world's easiest chore; there was a biochemical chasm between a molecule that could cut itself apart and one that could deftly weave nucleotides into a precise copy of itself. That chasm yawned all the wider for the fact that, despite all the celebrated tools and progress of genetic engineering and molecular biology, biochemists have surprisingly little insight into how complex molecules like RNA perform specific functions, let alone how to tailor them into the kind of chemical tour de force that Szostak was proposing. "The general feeling was that what I was trying to do was impossible," says Szostak. "But I was optimistic. I really did believe that the origin of life happened this way, and that if it did I should be able to do it in the lab."

Working out of a cramped office in the maze-like interior of the drab Boston building that serves as Massachusetts General Hospital's research facilities, Szostak adopted a working philosophy that was dazzlingly dull: think small. "If you looked at this goal in terms of a big jump, it really did seem impossible," he explains. "But if you looked at it as the sum of a lot of little, manageable steps, it didn't seem as hard." Step one for Szostak and graduate student Jennifer Doudna was to take Cech's *Tetrahymena* RNA enzyme, which boasts the ability to break itself apart, and get it to perform its chopping trick on a separate molecule, since an RNA molecule serving as a replication enzyme would have to act not on itself but on an identical, second RNA molecule. One possibility would be to break the *Tetrahymena* RNA into two separate pieces, and have one piece act on the other. But that would work only if the enzymatic splitting action could be localized to a piece of the strand—an open question, since that issue hadn't been particularly relevant to Cech and other researchers. Szostak and Doudna soon discovered that, in fact, the splitting action was due to a particular large chunk of the strand, comprising about 360 of the 400 nucleotides in the strand; the remaining, smaller chunk merely consisted of the target site for the splitting. Synthesizing the two chunks separately in a test tube, they confirmed that the larger chunk on its own did indeed act as an enzyme to tear apart the smaller, target chunk.

Even better, they were able to make the reaction run in the other direction: not only could the enzyme break the target in two, but it could also weld, or "ligate," the two broken pieces of the target. Thus the team had come up with an RNA enzyme that could convert two small strings, or "oligos," of about twenty nucleotides each into a single, longer strand. "It was a very important, very exciting first step," says Szostak.

Next, Szostak and Doudna had to find out if the enzyme would work with the right configurations of nucleotides. Of the four types of nucleotides in RNA, adenine prefers to pair up with uracil, while guanine matches up with cytosine; these two happy couples are called "Watson–Crick" pairs, while other, less felicitous combinations are called "wobble" pairs. When an enzyme grabs a template strand of RNA and builds a complementary strand alongside it, all it has to do to get it right is make sure that each nucleotide is lined up next to its Watson–Crick pal, so that there are no wobble pairs. But the *Tetrahymena* RNA enzyme seemed determined to deal only with oligos that ended in wobble pairs, a preference that would prevent it from accurately copying long chains of wobble-less RNA.

To remedy the situation, the two scientists looked for a way to subtly modify the properties of the enzyme just enough to get it to work with oligos that were entirely Watson–Crick-ish. But given enzymes' complexity, this sort of fine-tuning is always a hit-or-miss proposition; Szostak and Doudna decided to simply throw various chemicals at the enzyme and hope one of them did the trick. One of the first substances they tried was spermidine, a small, electrically charged molecule that had been shown in other labs' experiments to twist the molecular structure of enzymes. It was a shot in the dark, but luck was on their side: when spermidine was added to the RNA mixture, the enzyme readily ligated oligos without wobble-pair endings.

Pressing their good fortune, Szostak and Doudna then tried to get the enzyme to ligate not just oligos from the small chunk of *Tetrahymena* RNA, but a variety of oligos. Again, the enzyme worked perfectly, joining together as many as five different oligos into a single strand. More important, the RNA enzyme lined the five pieces up alongside a second strand identical to itself, so that each piece was complementary to the nucleotides of the second strand. In other words, the RNA was serving as both an enzyme and a template, fashioning complementary copies of itself. By 1988, the two were sure they were hot on the trail of self-copying RNA. "This was the first thing that really looked replication-like," says Szostak.

In fact, what they had come up with fell short of self-replication on one major criterion: their enzyme could only assemble long oligos, and not individual nucleotides, according to a template. That's a critical difference, because an RNA enzyme making copies out of such long, prefabricated oligos wouldn't have an opportunity to create the subtle variations that are crucial to the process of evolution. "The system was really only giving us back a longer version of whatever we put in," says Szostak. "We wanted to get it down to handling two- or three-nucleotide oligos, because then we could put in all the random combinations and let it pick the right ones." But the enzyme simply didn't have anywhere near the precision necessary to match up and ligate the 120 three-nucleotide oligos it would have needed to construct a copy of itself. Szostak and Doudna had to do one of two things: reengineer the enzyme to vastly improve its ability to handle a large number of oligos; or find an RNA enzyme that was short enough to be constructed from a relatively small number of oligos.

Szostak decided to take the second route. Looking over the fifty or so other RNA enzymes that had been identified since they had started their research, Szostak and Doudna settled on a strand of RNA called "sunY" found in a bacterial virus known as T4. SunY had one overriding grace: at about 200 nucleotides, it was the shortest RNA enzyme yet discovered.

But though it was half the length of the *Tetrahymena*, and even better at ligating oligos, sunY was still too long. It would never be able to reliably string together the nearly 70 three-nucleotide oligos it would need to replicate itself; it would be like presenting a child with a jigsaw puzzle of too many pieces. To improve the situation, Szostak and Doudna relentlessly hacked away nucleotides from the enzyme to see how far they could shrink it—in effect, simplifying the jigsaw puzzle. But it was a delicate game of trade-offs: while a shorter sunY would make an easier-to-replicate template, the molecule needed long "arms" to grab and ligate oligos. If they hacked off too much, sunY would no longer be an effective enzyme.

The two researchers eventually got down to a 160-nucleotide version of sunY. At first it proved too weak as an enzyme, but they discovered they could restore its enzymatic strength by shuffling around a few of the nucleotides. Still, after all this inspired tweaking, the resulting enzyme was much too large. "Every time we had come to a roadblock Jennifer and I would sit and talk about it and come up with some strategy to get around it," says Szostak. "It seemed amazing that we had gotten this far. But this was a severe problem, and we were very discouraged. We were forced to sit back and think about whether there was some completely different way to approach it."

They needed a breakthrough—and it wasn't long in coming. Szostak was sitting in his office staring at sunY molecular structure diagrams, when he suddenly had an idea. What if they could split the enzyme into three pieces that self-assembled? He ran out of the office to tell Doudna, who immediately got to work on the approach.

It was an entirely novel strategy that seemed to meet the team's two conflicting needs: RNA that was long enough to act as an enzyme, and short enough to be relatively easily copied. In theory, the three sunY "subunits" would link up to form the enzyme, grab an as-yet-unattached piece to serve as a template, and assemble oligos into a complementary strand alongside the template. Just two weeks after Szostak had come up with the idea, he and Doudna were watching it work on a bench. "It was one of the most exciting moments of the project," he says.

In a very rough sense, their three-piece RNA enzyme is indeed self-replicating. They still, however, have to feed it prefab oligos of about eight nucleotides each; until the system makes accurate copies out of randomly clumped two- or three-nucleotide chunks, it couldn't be considered a plausible model for what nature did 4 billion years ago. Right now, however, the enzyme is far too sloppy to stitch enough tiny oligos together: because it deviates from the template about one out of every three times it tacks on an oligo, the chances of its being able to get twenty right is minute. "We really need to focus on this question right now," says Szostak. "We have to go from 70 percent accuracy to 99 percent accuracy."

If he can get anywhere near that sort of accuracy, Szostak should be able to sit back and let nature take over the design work. As the enzyme replicated itself, errors and all, sooner or later it would make a mistake that resulted in a better self-replicator. That "mutant" molecule would come to dominate, eventually producing other mutants that did even better. At that point, evolution will have taken over.

For now, however, Szostak has had to turn to a sort of hand-operated version of evolution: instead of letting the sunY replicate itself, Szostak synthesizes it in a test tube, purposely shuffling some of the nucleotides around in random fashion in the hopes that a handful of the resulting possible trillions of combinations turn out to be more efficient enzymes. To isolate these talented few, he has the entire batch ligate various molecules, and strains the mixture to pull out the ones that have created the longest strings. He then synthesizes a fresh batch of the winners, and repeats the entire process. "Getting to real evolution is still a distant goal for us," he says, "but I'm pretty optimistic."

In addition to his work on RNA, Szostak is also experimenting with "phospholipid membranes," a soapy substance that forms microscopic globules. When the phospholipid material is added to a test tube full of his RNA molecules in solution, the emerging globules trap tiny drops of liquid—along with the RNA molecules in the drops. Though that doesn't even approximate the complex way in which modern DNA directs the construction of its own cell membrane, Szostak thinks the globules can perform many of the functions of a real membrane. He has already found ways of getting these globules to "grow" by letting them combine with other globules, as well as to divide by squeezing them through porous materials. "A lot of people neglect the idea of compartmentalizing the RNA, but separating it from the rest of the world is crucial for evolution," he says. "You want the RNA replicating its own interesting mistakes instead of having them wander off in the solution."

The hope, of course, is that some of these "interesting mistakes" could push the RNA in the direction of more and more complex configurations, until ultimately a quasi-living cell would emerge. If intelligence is a goal, the cell would require neuron-like properties, including the ability to send, receive, and process signals, as well as the ability to connect itself up to others of the same type in useful ways.

While all this may seem a tall order, scientists are already experimenting with techniques for "directed evolution"—the ability to speed up and tailor the process of evolution in a test tube. One of the leaders of this new field, Manfred Eigen of the Max Planck Institute in Germany, has already developed the ability to evolve viruses in carefully controlled ways by passing them through tubes filled with various physical and chemical obstacles. Each obstacle is designed to stop all viruses except those that have through mutation acquired a particular desired characteristic; these mutations pass through and encounter the next obstacle, which leads to yet another favorable mutation. In this way, a customized virus is created through a series of mutations. "I call it my evolution machine," chuckles Eigen.

These are important first steps, but clearly the jump to evolving self-connecting, neuron-like cells is not a small one. On the other hand, as David Jefferson points out, getting evolution kicked off is the hard part; then the process tends to pick up speed. An evolved semiartificial neural network may not be right around the corner, but neither need it be many decades away.

Even if researchers succeed in evolving a semiartificial neural network with the capacity and interconnectedness of a human brain, would such an entity be *like* a human brain? Or more specifically, could it achieve consciousness? The question of whether a fabricated device could become conscious has long been hotly debated. But the possibility that such a device might be a semiartificial neural network rather than a cleverly programmed digital computer adds a new twist to the matter that has already motivated the revisiting of some strongly held opinions.

7. SEEDS OF COGNITION

> *This is not just an ad hoc, superficial objection to the idea of artificial consciousness. It is a reason for saying that no rational design process, working from first principles, would ever be likely to succeed.*
>
> —NICHOLAS HUMPHREY

There is no precise, widely agreed upon definition of consciousness, but most of us have an intuitive sense of what is meant by the term. Consciousness, or cognition, is a sort of awareness—of self, of interaction with the world, of thought processes taking place, and of our ability to at least partially control these processes. We also associate consciousness with an inner voice that expresses our high-level, deliberate thoughts, as well as with intentionality and emotions.

It is not a given that consciousness is a necessary or even a highly desirable property for an intelligent machine, whether artificial or semiartificial. The goal of artificial intelligence has been to produce useful, intelligent behavior from a machine, and not to create a machine that feels happy or bored, or that is capable of reflecting on its thought processes. Certainly there are aspects of intelligence often associated with consciousness that would prove useful in an intelligent machine: the ability to focus attention on important events and details, for example, or to reason in a clear fashion. But functional versions of these capabilities can—and have already been—programmed into machines, in one form or another. Stephen Grossberg's ART can focus its "attention," for example, and John McCarthy has long been turning out programs that can reason. Yet such systems are clearly not conscious.

Of course, if a machine were to be so skillfully programmed that its behavior were to some extent indistinguishable from that of a human being, the question of consciousness might be seen as irrelevant—or even unanswerable, since there is probably no way to know what a machine "thinks" or "feels." Yet even if the question of what a machine may or may not think to itself ultimately proves to be impractical, it is at the very least a fascinating question with profound philosophical implications. And it may turn out that consciousness is critical to producing human-like intelligence.

In either case, the question of the reproducibility of consciousness has, especially in recent years, given rise to a great deal of passionate argument. The positions staked out by the arguers can roughly be placed into four categories:

- The *quasi-metaphysical* view, which holds that consciousness cannot be ascribed to purely physical processes, and thus is inaccessible to even an arbitrarily advanced scientific assault
- The *physical/irreproducible view*, which allows that consciousness arises purely from physical processes in the brain, but argues that these processes are so complex or otherwise so far removed from our scientific understanding that there is no practical hope of duplicating them
- The *physical/reproducible* view, which provides for the possibility that the processes giving rise to consciousness can be understood and duplicated, though it is likely to be an exceedingly difficult task
- The *physical/trivial* view, which holds that there is nothing all that special about consciousness, and that a machine packed with enough intelligence will more or less automatically acquire consciousness somewhere along the way

Proponents of AI's "strong claim," which essentially asserts that a machine could become conscious, hew to either the "reproducible" or the "trivial" views, while the more cautious observers who draw the line at the "weak claim" generally adhere to the "quasi-metaphysical" or "irreproducible" views. (Not all views fit neatly into these categories, of course. Nicholas Humphrey, for example, doesn't rule out the possibility that a machine could acquire something roughly equivalent to human consciousness, but he insists it could never be *exactly like* human consciousness, which developed from unique evolutionary considerations. This opinion straddles the irreproducible and reproducible views, as well as the strong and weak claims.)

The quasi-metaphysical view dominated from the time Descartes argued for it in the seventeenth century through much of this century, and some philosophers still favor it. A very few scientists continue to adhere to it as well, most notably neurophysiologist John Eccles at the Max Planck Institute in Frankfurt, Germany, who proposes the existence of nonmaterial entities called "psychons" that impart ideas and feelings by interacting with the brain. For the most part, though, the quasi-metaphysical view isn't taken all that seriously by the scientific community in general, and the AI or neuroscience communities in particular, all of whom tend to believe that whatever consciousness is, the brain does not achieve it by stepping outside the bounds of the physical world.

The "trivial" view is also a relatively isolated one, restricted primarily to conventional AI researchers like Doug Lenat. Lenat and his colleagues have good reason to adhere to this point of view: since their careers are dedicated to writing computer programs that presumably embody the essence of intelligence, the idea that the highest level of intelligence might involve some sort of physical mechanisms outside the bounds of programming would be a tacit admission of the inherent limitations of their approach. To virtually everyone else, however, it seems almost self-evident that consciousness *is* a special quality that would not automatically fall into the lap of even the most highly sophisticated program.

Thus for many philosophers, most scientists, the vast majority of neuroscientists, and for nature-based AI researchers, the effort to resolve the question of consciousness comes down to a tug-of-war between the irreproducible and reproducible views. The challenge for both camps is to come up with plausible theories regarding the nature of the physical processes that give rise to consciousness; given the processes, it should be clear whether or not they could in principle be made to take place in an artificial device.

The opening shot in this contemporary battle was a loud one fired on behalf of irreproducibility by University of California at Berkeley philosopher John Searle in 1980. Searle's argument, which was intended not so much to suggest the processes that give rise to consciousness as to suggest which processes could not give rise to consciousness, took the form of the hypothetical "Chinese Room." In this scenario, one imagines a non-Chinese-speaking person sitting in a room with a long list of rules for translating strings of Chinese characters into different strings of Chinese characters. When a string of characters is slipped under the door, the person consults the rules, produces a new string, and slips the results back out under the door. If the incoming strings actually comprised questions, then a particularly clever and exhaustive set of rules could conceivably allow the person in the room to produce outgoing strings that comprised answers to the questions.

From the point of view of a Chinese-speaking person outside the room slipping in questions, the room would seem to contain an intelligent person who is reading the questions and coming up with answers. And yet the person in the room has no idea what he or she is reading, nor what he or she is writing; the questions and answers are a meaningless gibberish of symbols. And this, said Searle, is the most that AI could hope to produce: a machine that produced intelligent-sounding answers without ever comprehending the meaning of anything. Or to put it another way, no matter how sophisticated a machine's programming, the machine could never be said to be conscious. It would be smart in the stupidest possible way.

Some conventional AI researchers shot back that a highly sophisticated AI program would not simply churn through a list of rules in a simplistic fashion, but rather would consider many different types of rules in parallel, deal with conflicts between rules, make guesses about rules, recognize relationships between rules, and construct new rules, much as Cyc does. Just as an especially astute person locked in the Chinese Room might eventually begin to understand Chinese, they suggested, so might a sophisticated rule-based system acquire the foundations of consciousness. Neural network researchers had more fertile ground over which to take Searle to task: Searle's Chinese Room was analogous to a single neuron, they said, firing according to a set of electrochemical "rules" without reflecting on the meaning of its signals. In the brain, consciousness arises from the interaction of many Chinese-Room-like neurons, and the same might happen in an artificial neural network.

Since introducing the Chinese Room, Searle has expanded his attack on AI to suggest that consciousness can never be simulated on any computing device, including an artificial neural network, because the brain itself doesn't perform anything resembling information processing. Instead, he notes, the brain simply provides for the carrying out of various biochemical interactions, without hinting at anything resembling programming. Since consciousness arises from these programless, noncomputational biochemical processes, he maintains, there is little hope that an artificial device could achieve consciousness.

But again, Searle's point seems fairly easy to counter. As Michael Conrad has shown in theory, and Robert Birge and Masuo Aizawa have shown experimentally, the line between biochemical interactions and information processing is a blurred one, and the next few years will almost certainly see the advent of rudimentary but entirely biochemical computing devices. By the same token, nonbiochemical neural networks have had great success in simulating the results of biochemical interactions, including many aspects of the brain's functioning.

The consciousness-as-neural-network notion, and with it the reproducibility camp, were given a prominent boost by Tufts University philosopher Daniel Dennett in his controversial 1991 book, *Consciousness Explained*. Dennett proposed a model for the mind in which multiple streams of signals simultaneously course through the brain's neural network. Each of these streams represents a sort of protothought that has not broken through to consciousness; various simultaneous protothoughts might represent complementary or conflicting views of some element of reality. As the brain continues to process information and gather new data, it eliminates some of these "drafts" of consciousness and consolidates others, finally weaving together out of this mishmash of thought patterns a single, complete, and well-integrated pattern that emerges as conscious thought. As a simplified example, imagine a driver glancing up and seeing something darting across the road ahead; one protothought might start to assemble an interpretation of the image as that of a squirrel, a second might be leaning toward a cat, and a third a shadow of a bird flying overhead. As information about the size, speed, and coloring of the object is taken into account, the driver's brain is quickly able to settle on the cat, and that draft becomes a conscious image.

Dennett doesn't try to specify exactly what physical mechanisms provide the brain with the ability to pull integrated, conscious thought out of a swarm of protothoughts. But, as Dennett recognizes, the concept of interacting patterns of signals settling into a stable pattern phenomenon clearly has a neural-network feel to it. Could an artificial neural network, then, achieve something like consciousness in the same fashion?

Existing neural networks have already displayed properties that are similar to those associated with certain simple elements of consciousness. Stephen Grossberg, of course, maintains his ART neural network provides a vague imitation of conscious perception in the way it mulls over possible categorization and then locks onto a single one in a "whoomph" of recognition. And he maintains that a more exact simulation of consciousness could be achieved by a more complex ART-like neural network organized in a hierarchical fashion, so each level makes a more sophisticated categorization of the patterns in the level below it. In this model, raw data would be processed in the lowest levels; this patterned raw data could be fashioned into fragments of thought in lower-middle levels; the fragments might be assembled into protothoughts in the upper-middle levels; and the protothoughts could be integrated into something akin to consciousness in the highest levels. What's more, says Grossberg, each level would self-organize as it gained experience in categorizing the patterns in the level below it, so that consciousness would emerge over time, much as it does with infants.

The lip-reading neural network developed by Stanford researcher

David Stork provides a surprising example of how neural networks can offer glimmers of the properties of consciousness. There is a well-known phenomenon known as the "McGurk effect" in which a person listening to the sound "ba," and simultaneously watching a person silently mouth the sound "ga," clearly hears the sound "da." The generally accepted explanation for this odd effect is that the brain takes the two conflicting pieces of incoming data into account, and reconciles them into a single sound that presents a good compromise. ("Da" is halfway between "ba" and "ga," from a vocal mechanics point of view.) Some researchers have suggested this phenomenon is an example of the brain weaving conflicting protothoughts into a single conscious perception. Yet when Stork presented his lip-reading neural network (which also takes audio data) with the same "ba/ga" input, it, too, came up with "da." "It's an interesting psychological effect," says Stork, "and our neural network seems to experience much the same thing."

But such effects, provocative as they may be, seem hardly to scratch the surface of consciousness. In fact, most neural network researchers suspect neural networks will require special properties to achieve consciousness. Some point to chaos as a potential missing ingredient. Chaotic systems are typically characterized by a tendency to suddenly jump from irregular, seemingly random behavior to stable, orderly behavior, just as water going down a drain in a messy gurgle can suddenly form an orderly funnel. Perhaps this schizophrenic behavior could provide a physical dividing line between conscious and non-conscious intelligence: conscious thought could be the result of disordered signal patterns settling into stable modes. One appealing aspect of a chaotic consciousness is the provision for unpredictability and variability in the resulting thoughts. Unlike standard artificial neural networks, the brain isn't constrained to producing the same results every time it processes the same information under the same conditions. Neither would a chaotic neural network, as Kazuyuki Aihara demonstrated with his chaotic memory network. As a further incentive, studies of brainwaves have clearly shown that the brain is loaded with chaos.

Another physical mechanism that could provide a threshold for conscious thought, and one that's currently receiving a great deal of attention from neuroscientists, is correlations in the timing and rate of the firing of different neurons. Studies have shown that when animals perform certain challenging tasks, large groups of neurons tend to fire either at roughly the same instant, or else with the same frequency, typically about forty times a second. Researchers speculate that conscious thought may consist of those thought patterns represented by these waves of neural synchronization, while nonconscious thought occurs in less synchronized fashion. Work has already begun on setting up artificial neural networks that incorporate synchronized and unsynchronized firing patterns.

In fact, artificial neural networks have proven, at least in principle, entirely well suited to simulating all of the various physical mechanisms that have been proposed as possible sources of consciousness in the brain. All, that is, except for the one proposed by Roger Penrose, the celebrated Oxford physicist who has in recent years made a not entirely welcome excursion into AI.

Penrose's 1989 book, *The Emperor's New Mind*, rocked the AI community harder than did Searle's Chinese Room. The message was essentially the same: a computer could never be conscious, and thus truly intelligent. But whereas Searle employed a simple metaphor to make his case, Penrose wove elements of mathematical philosophy, information science, cognitive psychology, and physics into a lengthy, detailed, tightly constructed (and yet somehow still readable) argument that left AI researchers gritting their teeth but at something of a loss as to how to counterattack. If there were gaping holes in his argument, they were buried somewhere in or behind the many technical points Penrose threw out at every turn. As a result, most AI researchers have just shrugged their shoulders, assumed Penrose missed the boat somewhere along the way, and gone on with their work. Particularly irked have been nature-based AI researchers, who assert Penrose was inappropriately lumping their efforts in with those of traditional AI researchers. "Penrose was basing his arguments on intuitions coming from what AI has been about, rather than on what AI is becoming," grumbles Santa Fe Institute researcher Melanie Mitchell, who evolves cellular automatons.

Penrose's arguments don't readily lend themselves to summary, but one of his key assertions is that consciousness requires some special physical mechanism to achieve the Dennett-like juggling act of dealing with multiple, simultaneous patterns of protothoughts before zeroing in on one unified pattern that becomes conscious thought. This mechanism, he claims, would have to have "nonlocal" properties—a physicist's way of saying that some aspects of thought patterns would have to act more or less instantly across widely separated locations of the brain, rather than being limited to spreading out relatively slowly in neuron-by-neuron fashion.

Though nonlocality may sound like an implausible concept, it is a property that turns up in many aspects of nature. For example, materials that crystallize—that is, whose normally disorganized atoms can suddenly rearrange themselves into a well-ordered pattern, as when water turns to ice—often do so in nonlocal fashion, so that a consistent pattern of crystallization appears simultaneously in different areas of the material. But nature's most fundamental and intriguing display of nonlocality occurs in quantum mechanics, the strange but exhaustively tested theory at the heart of modern physics. And it is here that Penrose believes the secret ingredient of consciousness lies.

According to quantum mechanics (or at least to its conventional interpretation), any bit of matter or energy is at first a wave, and then at some point is transformed into a particle. As particles, matter and energy behave more or less as we would expect little billiard balls to act. But the picture is weirder when a drop of matter or energy is in a wave state: it must roughly be thought of as simultaneously existing in a virtually infinite number of locations, and with each of its properties—energy, spin, sometimes even mass—taking on multiple potential values. Quantum mechanics reconciles this twilight zone of indeterminacy with the ordinary world we see around us by proclaiming that the act of observation somehow causes matter and energy to revert to the particle state—that is, the quantum mechanical wave "collapses." What's more, the wave state collapse is entirely nonlocal: when one part of a wave collapses, any other part of it or any wave that is closely related to it will also collapse at exactly the same moment, no matter how far away.

The question of exactly how observation causes matter and energy to make the transition from wave weirdness to well-behaved particle is one that most physicists don't even ask—the theory works, so why question it?—and those that do ask don't all agree on an answer. Penrose has his own controversial ideas about the collapse of the wave state, relating to the influence of gravity. But these issues aside, Penrose asserts that the wave/particle duality of matter and energy offers a perfect foundation for understanding consciousness. Before a thought—or the neural signals that constitute thought—enters consciousness, he says, it exists in a quantum wave state, and thus perhaps actually comprises any number of different, simultaneous protothoughts. At the threshold of consciousness, the wave-thoughts might then collapse into a single ordinary thought.

If this or other quantum mechanical phenomena are indeed behind consciousness, then the problem for AI is obvious: existing computing devices, including artificial neural networks, may not be able to simulate them. Largely on these grounds, Penrose proclaims AI—or at least that segment of AI that hopes to achieve human-like intelligence—to be a dead end.

Since the appearance of *The Emperor's New Mind*, and the resulting firestorm of resentment it drew from the AI community, Penrose has been busy shoring up and expanding his assertions, hewing to much the same line. Only one large gap remained in his argument, he felt: How do neurons provide the necessary quantum mechanical mechanisms? Since the phenomenon would be occurring at a molecular or even atomic scale, it wasn't surprising that neuroscientists hadn't discovered it yet; indeed, it isn't even clear what it is they would be looking for, since physicists don't understand quantum collapse. But a clue as to the nature of the mechanism would have left him feeling more assured. "We needed something in the brain that could produce coherent quantum effects over a large area," he recalls.

The clue came from the unlikely source of Stuart Hameroff, the University of Arizona anesthesiologist who studies the submicroscopic "microtubules" that provide internal physical structure, and possibly an internal computer, to neurons. Hameroff and other scientists working with microtubules had demonstrated that microtubule networks seemed to have nonlocal properties: introducing an electric signal at one point in the network could almost instantly alter the entire network, a property that had already suggested to Hameroff a possible mechanism for consciousness. What's more, the process by which microtubules "switched"—the bouncing back and forth of a single electron—appeared to be dominated by quantum mechanical effects. (Quantum mechanics tends to have negligible effects on large objects like baseballs, but huge effects on tiny objects like electrons.) Thus the propagation of a signal through a microtubule network would have to be regarded as a quantum mechanical phenomenon.

Recognizing that the conceptually peculiar and mathematically abstruse discipline of quantum mechanics was somewhat over his head, Hameroff decided to consult someone well versed in that field who also happened to be interested in the phenomenon of thought. That person, of course, was Roger Penrose. "I had always thought the weakest part of Penrose's argument was trying to relate quantum effects to single neurons," Hameroff says. "It makes much more sense to do it at the level of single particles, which is the way microtubules work."

Hameroff's quest could easily have turned out to be the scientific version of *Roger and Me*; after all, one might expect a renowned physicist and AI skeptic like Penrose to look somewhat askance at the musings of an anesthesiologist who talks about building self-organizing nanorobots and altering consciousness via microwaves. But Penrose, who is nothing if not intellectually adventurous, gave Hameroff an audience. What he heard intrigued him, and the more he thought about it the more he liked it. "His ideas were a bit different," says Penrose. "But here was a place in the brain to look for quantum effects." Two weeks after meeting with Hameroff, Penrose was telling a Cambridge University audience that microtubules were his leading candidate for the seat of consciousness.

Hameroff, of course, believes a semiartificially intelligent device can be constructed by genetically engineering microtubules that self-organize into networks. Penrose, in what could be taken as a surprising concession in his otherwise four-square stand against AI, doesn't rule it out. "I don't say one way or the other whether it's possible," says Penrose. "We'd have to learn more about the physics involved. But I don't yet see any reason why it wouldn't be allowed."

•   •   •

Nature may have found a way to employ quantum wave/particle du-ality in brains, but doing the same in an artificial neural network is another matter. Since quantum effects are generally limited to the smallest size scales, such a network would have to function at the level of individual atoms and electrons. Or would it?

IBM researcher Claudia Tesche, backed by theoretical work from Tony Leggett at the University of Illinois at Urbana-Champaign, has been trying to construct a device in which a SQUID—the pinhead-sized semiconductor component used to detect magnetic fields—ex-hibits quantum duality even though it is a trillion trillion times larger than an atom. In behaving like a wave, the SQUID would allow a tiny current to run around it in both clockwise and counterclockwise di-rections simultaneously; when the wave collapsed, the current would choose one of the directions. Such behavior in a macroscopic object would violate the conventional interpretation of quantum mechan-ics, but Tesche and Leggett believe this interpretation is wrong, and believe the SQUID experiment will prove it.

If a SQUID were, in fact, able to exhibit such quantum duality, then one could conceive of a network of SQUIDs carrying multiple signal patterns in the wave state before collapsing to a single pattern, much as Penrose suggests microtubule networks do in the brain. But Tesche, who also uses SQUIDs to study brain activity, notes her ex-periment is in the early stages, and even if it works she has no plans to steer it toward quantum neural networks.

Penrose, for his part, rules out any scheme to imitate consciousness out of totally artificial devices. Even Hameroff is skeptical of build-ing brain-like entities based on quantum-mechanical semiconductor technology, noting that microtubules' cylindrical shape traps and fo-cuses quantum mechanical waves in a way that is probably crucial to the microtubules' functioning. "You could never get the right mole-cular architecture from silicon," he contends. "There is something very special, almost magical, about microtubules." Hameroff does al-low, however, that scientists might someday synthesize a different sort of nonorganic molecule with some of the properties of microtubules. Buckminsterfullerine, or "buckyballs," the recently discovered near-spherical molecule consisting of sixty carbon atoms, may even be a candidate, he speculates.

• • •

But why look for an artificial version of microtubules, or any other biological component, if the original or something like it can be made available through biotechnology, or perhaps through directed evolution? One of the main points of the nature-based AI movement is to eliminate the traditional emphasis on building from scratch, and instead to take advantage of whatever shortcuts nature offers.

As the nature-based AI movement continues to thrive, and the conventional AI effort continues to falter, it seems increasingly likely that the first conscious device will not only provide the functionality of a brain, it will in many ways be *like* a brain. Nature has provided a blueprint for intelligence, and AI is finally prepared to follow it.

CONCLUSION

The nature-based AI movement seems exceptionally well positioned to succeed. The fields of science on which the movement is most dependent—molecular biology, neuroscience, and complex adaptive systems—are the very fields currently making the greatest progress. As it integrates successes in these fields and builds upon them, nature-based AI could very well reach a critical mass in which advancement simply continues to accelerate, so that the construction of a semiartificial intelligence would occur perhaps in a decade or two.

On the other hand, conventional AI also got off to a breathtaking start, building on the then rapidly growing fields of semiconductor technology, cognitive psychology, and information science, among others. It wasn't for nearly two decades that the endeavor began to lose momentum.

But there are reasons to believe that nature-based AI will remain on the fast track. For one thing, neural network research, despite spending some two decades out of favor for the wrong reasons, has proven to be the single most enduring and widely followed approach to AI. For another, nature-based AI's emphasis on complex interactions and self-organization over conventional AI's reductionism and analytical purity is increasingly mirrored not only in many other areas of science but even in politics and economics. Communism and its top-down, analytical machinery, for example, have been widely rejected in favor of the unpredictable and yet effective "invisible hand" that emerges from the seemingly chaotic interactions of the free marketplace and electorate. Even U.S. corporations and the military, two traditional bastions of centralized control, are being slowly pushed to more bottom-up structures where some of the most important decisions are made not by CEOs and generals but by "workgroups" and platoons. It's possible that these changes represent some sort of large-scale fad, but they have the feeling of a long-term evolutionary shift. If so, nature-based AI is on the right side of the fence. Besides, in these times of rapidly growing respect for nature's achievements, any approach that takes nature for its closely held role model seems all the more well considered.

Even if nature-based AI turns out to be on the right path to building a semiartificial brain, do we want such an entity? As Roger Schank points out, we already have plenty of human brains at our disposal.

On a practical level, one could respond by pointing out that there are many human brains that would profit from, and be cheered by, the opportunity to pass off to a device some of the less interesting tasks that currently consume them, just as calculators usefully freed up the time of millions of accountants and clerks. But such arguments are really beside the point. There is no way of predicting the impact on society of truly intelligent devices. The one way to settle the question is to accept the challenge of building them.

As MIT researcher Pattie Maes puts it: "It is the role of the scientist to push for extremes."

NOTES ON SOURCES

Most of the information about specific researchers in this book came from interviews—usually in person, but sometimes on the phone—with the researchers themselves and with their close associates. I thank them for their time and apologize that, as is usually the case, only a fraction of what they told me has ended up in print. In cases where researchers are mentioned briefly, I've sometimes relied primarily on their technical papers and books, and on descriptions of their work by other researchers. Some books and articles by other journalists proved helpful as background material, and in a very few instances as sources for facts appearing in this book; I've tried to identify these cases in the following further reading list, and I apologize if any have slipped through unnoted.

SUGGESTED FURTHER READING

The following list of books and articles represents a recommended, generally nontechnical, sampling of the enormous array of material available on artificial intelligence and the many related topics covered in this book.

Books

Adaptation in Natural and Artificial Systems, John H. Holland, Ann Arbor, Michigan: University of Michigan Press, 1975. This book helped lay the foundation for the artificial life movement.

Apprentices of Wonder: Inside the Neural Network Revolution, William F. Allman, New York: Bantam Books, 1989. A chronological account of the development of neural networks, including a superb easy-to-follow explanation of how neural networks function. Allman's book was one of the sources for the section of chapter 3 of this book dealing with the history of neural networks.

Artificial Intelligence, Patrick Henry Winston, Reading, Massachusetts: Addison-Wesley, 1979. One of the best-known introductory textbooks on the subject.

Artificial Intelligence at MIT: Expanding Frontiers (two volumes), edited by Patrick Henry Winston with Sarah Alexandra Shellard, Cambridge, Massachusetts: The MIT Press, 1990. A loosely related collection of technical and semitechnical papers that provides an excellent means for assessing AI's progress at one of the world's leading labs.

The Artificial Intelligence Debate: False Starts, Real Foundations, edited by Stephen R. Graubard, Cambridge, Massachusetts: The MIT Press, 1988. This outstanding, generally readable collection of essays examines some of the triumphs and failures of conventional AI through the eyes of many of its participants.

Artificial Life: The Quest for New Creation, Steven Levy, New York: Pantheon, 1992. A richly detailed journalistic account of the birth and progress of this strange and exciting field, a field that occasionally overlaps with the approaches to AI discussed in this book.

The Biology of the Brain: From Neurons to Networks, edited by Rodolfo R. Llinas, New York: W. H. Freeman, 1989. Part of the "Readings from *Scientific American*" series, this collection of previously published articles lays out the foundations and breakthroughs of neuroscience in that magazine's clear, if sometimes slightly dense, style. It served as one of the sources for the information about neurons in chapter 4 of this book.

Complexity, M. Mitchell Waldrop, New York: Simon & Schuster, 1992. A captivating blow-by-blow description of the early goings on at the Santa Fe Institute, the eccentric but high-powered scientific think tank that has become the leading center for studies of the emergent behavior of complex systems, including artificial life.

The Computer and the Mind: An Introduction to Cognitive Science, P. N. Johnson-Laird, Cambridge, Massachusetts: Harvard University Press, 1988. An exploration of some of the basic and not-so-basic issues in both high- and low-level intelligence.

Consciousness and the Computational Mind, Ray Jackendoff, Cambridge, Massachusetts: The MIT Press, 1987. Examines certain aspects of consciousness from a brain-as-computing-device point of view.

Designing Autonomous Agents: Theory and Practice from Biology to Engineering and Back, edited by Pattie Maes, Cambridge, Massachusetts: The MIT Press, 1991. A series of fairly technical papers on a generally nature-based facet of AI.

The Emperor's New Mind: Concerning Computers, Minds, and the Laws of Physics, Roger Penrose, New York: Oxford University Press, 1989. Not only a hard-hitting argument for a provocative new theory of the nature of intelligence and the limits of conventional AI but also a brilliant primer on physics, mathematics, computers, information theory, and neuroscience.

From Animals to Animats: Proceedings of the International Conference on Simulation of Adaptive Behavior, edited by Jean-Arcady Meyer and Stewart W. Wilson, Cambridge, Massachusetts: The MIT Press/Bradford Books, 1991. Papers from the conference that largely solidified the nature-based AI movement. Unfortunately for the average reader, many of these articles are somewhat technical.

The Hedonistic Neuron: A Theory of Memory, Learning, and Intelligence, A. Harry Klopf, New York: Hemisphere Publishing, 1982. A somewhat off-beat premise, but plausible and well-supported.

In the Image of the Brain: Breaking the Barrier Between the Human Mind and Intelligent Machines, Jim Jubak, Boston: Little, Brown, 1992. A detailed, entertaining look at the growing overlap between the fields of neuroscience and neural networks.

In Our Own Image: Building an Artificial Person, Maureen Caudill, New York: Oxford University Press, 1992. A readable, fairly detailed look at a few of the key challenges in AI, including machine vision, speech recognition, and robotic motion. Served as a source for chapter 3 of this book.

Mind Children, Hans Moravec, Cambridge, Massachusetts: Harvard University Press, 1988. An imaginative and very readable account by a celebrated roboticist of the direction in which robots might be heading.

Neurophilosophy: Toward a Unified Science of the Mind-Brain, Patricia Smith Churchland, Cambridge, Massachusetts: The MIT Press, 1986. A challenging weave of philosophy and science.

Robotics in Service, Joseph F. Engelberger, London: Kogan Page Ltd., 1989. An overview of primarily industrial robotic applications.

Robots: The Quest for Living Machines, Geoff Simons, London: Cassell, 1992. A readable overview of some of the conventional and leading-edge approaches in robotics.

Tell Me a Story: A New Look at Real and Artificial Memory, Roger C. Schank, New York: Charles Scribner's Sons, 1990. An entertaining and personal, if somewhat eccentric, take on the nature of intelligence.

Vehicles: Experiments in Synthetic Psychology, Valentino Braitenberg, Cambridge, Massachusetts: The MIT Press, 1984. Simple toy cars exhibit increasingly sophisticated emergent behaviors in this delightfully bizarre book.

What Is Life? (available in an edition that includes *Mind and Matter* and autobiographical sketches), Erwin Schrodinger, Cambridge, England: Cambridge University Press, first published in 1944. Not an easy read, but a unique, thought-provoking perspective from one of the pioneers of quantum mechanics.

Articles (grouped by subjects roughly corresponding to the chapters of this book, and listed in alphabetical order within each group):

Robotics:
Computer, June 1989. A special issue on robots.

"For the Love of Robotics," John Schwartz with Alden Cohen, *Newsweek,* March 9, 1992, pp. 68–69. A brief look at some offbeat approaches to robotics. Source for the robotics market size estimate in this book.

"Go Robots, Go!" Judith Anne Yeaple, *Popular Science*, December 1992, starts on p. 96. The cleverly dumb robotic winners in a maze-navigating contest.

"In Search of the Human Touch," Ivan Amato, *Science*, November 27, 1992, pp. 1436–37. Efforts to give robot hands a sense of touch.

"The Inferno Revisited," Richard Monastersky, *Science News*, June 6, 1992, pp. 376–78. About Carnegie-Mellon's volcano-exploring robot, Dante.

"Rover on a Chip," Rodney A. Brooks and Anita M. Flynn, *Aerospace America*, October 1989, pp. 22–26. The scheme for tiny Mars robot explorers, in the designers' own words.

"Screen Robots Tell a Tale of Mankind," Thomas Hine, *The New York Times*, October 3, 1991. Hollywood's changing view of robotics and its cultural sources.

"Silicon Babies," Paul Wallich, *Scientific American*, December 1991, pp. 124–34. An overview of new approaches to robot control.

Conventional AI:

"CYC: A Mid-Term Report," R.V. Guha and Douglas B. Lenat, *AI Magazine*, Fall 1990, pp. 32-59. A lengthy description of CYC's goals and progress written by the program's creators.

"John McCarthy: Approaches to Artificial Intelligence," Reid G. Hoffman, *IEEE Expert*, June 1990, pp. 87-89. An interview with the father of the logic approach to AI.

"Northwestern's New A.I. Hotshot," Harold Henderson, [Chicago] *Reader*, December 15, 1989. A look at Roger Schank and his new department. Source for Schank's "Medicis" quote in this book.

"The Quest for the Thinking Computer," Robert Epstein, *AI Mag-*

azine, Summer 1992, pp. 81-95. Discusses the annual Turing Test competition, and provides excerpts.

"The World's Next Chess Champion?" Dawn Stover, *Popular Science,* March 1991, starts on p. 68. A look at the history of computer-playing chess programs. Source for some of the facts on chess-playing computers that appeared in this book.

Neural Networks:

"Going Beyond AI," Mickey Williamson, *CIO,* January 1992, pp. 62-65. Describes the basics of neural networks, and discusses their current and potential applications in business.

Intelligence: The Future of Computing, edited by Edward Rosenfeld, New York. A newsletter about new computing technologies, including neural networks.

"Neural Nets Tell Why," Casimir C. Klimasauskas, *Dr. Dobb's Journal,* April 1991, pp. 16-24. A somewhat technical look at the basic operation of a neural network, with detailed examples.

Scientific American, September, 1992. A special issue on the brain, including an article on how artificial neural networks work. Served as a source for this book.

Neurons and Artificial Neurons:

"Brain Flicks: Doctors Film an Obsession," David Stipp, *The Wall Street Journal,* December 2, 1992, starts on p. B1. A new technique for filming brain activity patterns.

"Japan Plans Computer to Mimic Human Brain," Andrew Pollack, *The New York Times,* August 25, 1992, starts on p. C1. Quoted from in this book.

"A Japanese 'Flop' That Became a Launching Pad," Neil Gross, *Business Week,* June 8, 1992. A comparison of Japan's fifth- and sixth- generation computer projects.

"Mapping the Brain," Sharon Begley with Lynda Wright, Vernon

Church and Mary Hager, *Newsweek*, April 20, 1992, pp. 66-70. An overview of new techniques for capturing images of the brain in action.

"The Mind and Donald O. Hebb," Peter M. Milner, *Scientific American*, January 1993, pp. 124-29. A mini-biography of the pioneer of neural learning.

"Optical Imaging Offers Gentler Way to Monitor Human Brain at Work," Warren E. Leary, *The New York Times*, August 25, 1992, p. C3.

"A Pioneer Is Out on a Limb Again," Lawrence M. Fisher, *The New York Times*, January 21, 1990, p. 8. Carver Mead and his determination to build chips that mimic groups of neurons.

"The Real World Computing Program: MITI's Next Computer Research Initiative," Akinori Yonezawa, *Science*, October 23, 1992, pp. 581-82. A discussion of the sixth generation project.

"What Is BioComputing?" Ray Valdes, *Dr. Dobb's Journal*, April 1991, starts on p. 46. Discusses comparisons between the brain and various computing technologies, and discusses attempts to bridge the gap.

Molecular Engineering:

Computer, November 1992. A special issue on molecular computing.

"Diminishing Dimensions," Elizabeth Corcoran, *Scientific American*, November 1990, pp. 122-31. An overview of the quest to shrink semiconductor technology to the molecular size scale.

"Engineering at the Lower Limits of Size," Anne Simon Moffat, *Mosaic*, Winter 1990, pp. 30-40. Various efforts to build microscopic structures.

"Molecular Computer Memory," D. Haarer, *Nature*, January 23, 1992, pp. 297-98. The basic considerations in trying to shrink computer components to the molecular level.

"Quantum Dots," Mark A. Reed, *Scientific American*, January 1993, pp. 118-23. A particular approach to ultra-tiny electronic building blocks.

"The Quest for the Molecular Computer," Mark A. Clarkson, *Byte*, May 1989, pp. 268-73. Looks at specific approaches to molecular computing.

Artificial Evolution:

"All the Way with RNA," Lori Oliwenstein, *Discover*, January 1993, p. 69. Briefly describes a breakthrough in the quest to link RNA to the emergence of life.

"Artificial Life: A Constructive Lower Bound for Artificial Intelligence," Richard K. Belew, *IEEE Expert*, February 1991, pp. 8-14. Discusses the overlap between A-life and AI.

"Directed Molecular Evolution," Gerald F. Joyce, *Scientific American*, December 1992, pp. 90-97. A detailed look at custom-evolved molecules by one of the leaders of the field.

"Duplicating Life—Mistakes and All," David L. Chandler, *The Boston Globe*, March 9, 1992. A look at self-duplicating-molecule creators Julius Rebek and Jack Szostak.

"Genetic Algorithms," John H. Holland, *Scientific American*, July 1992, pp. 66-72. An overview of this technique and its applications, by its founder.

"New Way to Develop High-Tech Drugs Monkeys with Darwin's Famed Theory," Jerry E. Bishop, *The Wall Street Journal*, February 25, 1993, starts on p. B1. How custom evolution can be applied to the search for new drugs.

Consciousness:

"Consciousness Raising," Bruce Bower, *Science News*, October 10, 1992, pp. 232-35 (part 2 appeared in the October 17 issue). Surveys the various ideas about how the mind arises from the brain.

FURTHER READING

"The Consciousness Wars," Robert K.J. Killheffer, *Omni*, October 1993, pp. 50-59. Discusses the debate over whether consciousness is a purely physical process.

"Is the Mind an Illusion?" David Gelman with Debra Rosenberg, Paul Kandell, and Rebecca Crandall, *Newsweek*, April 20, 1992, pp. 71-72. Also on the debate over consciousness.

"Nerve Cell Rhythm May Be Key to Consciousness," Sandra Blakeslee, *The New York Times*, October 27, 1992, starts on p. C1. Looks at some of the theories on how neurons conspire to create conscious perceptions. Served as a source for this book.

"The Problem of Consciousness," Francis Crick and Christof Koch, *Scientific American*, September 1992, pp. 153-59. Examines how experiments with human vision provide clues to how perception and consciousness arise in the brain.

INDEX

ABOUT THE AUTHOR

David H. Freedman is a contributing editor to *Discover* magazine and a regular contributor to *Science, Forbes ASAP, CIO,* and *Self.* He has also written for *The Boston Globe, The Boston Globe Magazine, The Washington Post, Inc.,* and *The Harvard Business Review.* He lives in Brookline, Massachusetts, with his wife and two children.